The Life Story of Insects

by Geo. H. Carpenter

PREFACE

The object of this little book is to afford an outline sketch of the facts and meaning of insect-transformations. Considerations of space forbid anything like an exhaustive treatment of so vast a subject, and some aspects of the question, the physiological for example, are almost neglected. Other books already published in this series, such as Dr Gordon Hewitt's _House-flies_ and Mr O H. Latter's Bees and Wasps, may be consulted with advantage for details of special insect life-stories. Recent researches have emphasised the practical importance to human society of entomological study, and insects will always be a source of delight to the lover of nature. This humble volume will best serve its object if its reading should lead fresh observers to the brookside and the woodland.

G.H.C.

DUBLIN,

July, 1913.

CONTENTS

CHAP.

CHAPTER I

INTRODUCTION

Among the manifold operations of living creatures few have more strongly impressed the casual observer or more deeply interested the thoughtful student than the transformations of insects. The schoolboy watches the tiny green caterpillars hatched from eggs laid on a cabbage leaf by the common white butterfly, or maybe rears successfully a batch of silkworms through the changes and chances of their lives, while the naturalist questions yet again the 'how' and 'why' of these common though wondrous life-stories, as he seeks to trace their course more fully than his predecessors knew.

Everyone is familiar with the main facts of such a life-story as that of a moth or butterfly. The form of the adult insect (fig. 1 _a_) is dominated by the wings--two pairs of scaly wings, carried respectively on the middle and hindmost of the three segments that make up the thorax or central region of the insect's body. Each of these three segments carries a pair of legs. In front of the thorax is the head on which the pair of long jointed feelers and the pair of large, sub-globular, compound eyes are the most prominent features. Below the head, however, may be seen, now coiled up like a watch-spring, now stretched out to draw the nectar from some scented blossom, the butterfly's sucking trunk or proboscis, situated between a pair of short hairy limbs or palps (fig. 2). These palps belong to the appendages of the hindmost segment of the head, appendages which in insects are modified to form a hind-lip or labium, bounding the mouth cavity below or behind. The proboscis is made up of the pair of jaw-appendages in front of the labium, the maxillae, as they are called. Behind the thorax is situated the _abdomen,_ made up of nine or ten recognisable segments, none of which carry limbs comparable to the walking legs, or to the jaws which are the modified limbs of the head-segments. The whole cuticle or outer covering of the body, formed (as is usual in the group of animals to which insects belong) of a horny (chitinous) secretion of the skin, is firm and hard, and densely covered with hairy or scaly outgrowths. Along the sides of the insect are a series of paired openings or spiracles, leading to a set of air-tubes which ramify throughout the body and carry oxygen directly to the tissues.

Such a butterfly as we have briefly sketched lays an egg on the leaf of some suitable food-plant, and there is hatched from it the well-known crawling larva[1] (fig. 1 _b, c, d_) called a caterpillar, offering in many superficial features a marked contrast to its parent. Except on the head, whose surface is hard and firm, the caterpillar's cuticle is as a rule thin and flexible, though it may carry a protective armature of closely set hairs, or strong sharp spines. The feelers (fig. 3 _At_) are very short and the eyes are small and simple. In connection with the mouth, there are present in front of the maxillae a pair of mandibles (fig. 3 _Mn_), strong jaws, adapted for biting solid food, which are absent from the adult butterfly, though well developed in cockroaches, dragon-flies, beetles, and many other insects. The three pairs of legs on the segments of the thorax are relatively short, and as many as five segments of the abdomen may carry short cylindrical limbs or pro-legs, which assist the clinging habits and worm-like locomotion of the caterpillar. No trace of wings is visible externally. The caterpillar, therefore, differs markedly from its parent in its outward structure, in its mode of progression, and in its manner of feeding; for while the butterfly sucks nectar or other liquid food, the caterpillar bites up and devours solid vegetable substances, such as the leaves of herbs or trees. It is well-known that between the close of its larval life and its attainment of perfection as a butterfly, the insect spends a period as a pupa (fig. 1 _e_) unable to move from place to place, and taking no food.

[1] The term larva is applied to any young animal which differs markedly from its parent.

Such, in brief, is the course of the most familiar of insect life-stories. For the student of the animal world as a whole, this familiar transformation raises some startling problems, which have been suggestively treated by F. Brauer (1869), L.C. Miall (1895), J. Lubbock (1874), R. Heymons (1907), P. Deegener (1909) and other writers[2]. To appreciate these problems is the first step towards learning the true meaning of the transformation.

[2] The dates in brackets after authors' names will facilitate reference to the Bibliography (pp. 124-8).

The butterfly's egg is absolutely and relatively of large size, and contains a considerable amount of yolk. As a rule we find that young animals hatched from such eggs resemble their parents rather closely and pass through no marked changes during their lives. A chicken, a crocodile, a dogfish, a cuttlefish, and a spider afford well-known examples of this rule. Land-animals, generally, produce young which are miniature copies of themselves, for example horses, dogs, and other mammals, snails and slugs, scorpions and earthworms. On the other hand, metamorphosis among animals is associated with eggs of small size, with aquatic habit, and with relatively low zoological rank. The young of a starfish, for example, has hardly a character in common with its parent, while a marine segmented worm and an oyster, unlike enough when adult, develop from closely similar larval forms. If we take a class of animals, the Crustacea, nearly allied to insects, we find that its more lowly members, such as 'water-fleas' and barnacles, pass through far more striking changes than its higher groups, such as lobsters and woodlice. But among the Insects, a class of predominantly terrestrial and aerial creatures producing large eggs, the highest groups undergo, as we shall see, the most profound changes. The life-story of the butterfly, then, well-known as it may be, furnishes a puzzling exception to some wide-reaching generalisations concerning animal development. And the student of science often finds that an exception to some rule is the key to a problem of the highest interest.

During many centuries naturalists have bent their energies to explain the difficulties presented by insect transformations. Aristotle, the first serious student of organised beings whose writings have been preserved for us, and William Harvey, the famous demonstrator of the mammalian blood circulation two thousand years later, agreed in regarding the pupa as a second egg. The egg laid by a butterfly had not, according to Harvey, enough store of food to provide for the building-up of a complex organism like the parent; only the imperfect larva could be produced from it. The larva was regarded as feeding voraciously for the purpose of acquiring a large store of nutritive material, after which it was believed to revert to the state of a second but far larger egg, the pupa, from which the winged insect could take origin. Others again, following de Rumur (1734), have speculated whether the development of pupa within larva, and of winged insect within pupa might not be explained as abnormal births. But a comparison of the transformation of butterflies with simpler insect life-stories will convince the enquirer that no such heroic theories as these are necessary. It will be realised that even the most profound transformation among insects can be explained as a special case of growth.

CHAPTER II

GROWTH AND CHANGE

The caterpillar differs markedly from the butterfly. As we pursue our studies of insect growth and transformation we shall find that in some cases the difference between young and adult is much greater--as for example between the maggot and the house-fly, in others far less--as between the young and full-grown grasshopper or plant-bug. It is evidently wise to begin a general survey of the subject with some of those simpler cases in which the differences between the young and adult insect are comparatively slight. We shall then be in a position to understand better the meaning of the more puzzling and complex cases in which the differences between the stages are profound.

In the first place it is necessary to realise that the changes which any insect passes through during its life-story are essentially accompaniments of its growth. The limits of this little book allow only slight reference to features of internal structure; we must be content, in the main, to deal with the outward form. But there is an important relation between this outward form and the underlying living tissues which must be clearly understood. Throughout the great race of animals--the Arthropoda--of which insects form a class, the body is covered outwardly by a cuticle or secretion of the underlying layer of living cells which form the outer skin or _epidermis_[3] (see fig. 10 ep, cu, p. 39). This cuticle has regions which are hard and firm, forming an exoskeleton, and, between these, areas which are relatively soft and flexible. The firm regions are commonly segmental in their arrangement, and the intervening flexible connections render possible accurate motions of the exoskeletal parts in relation to each other, the motions being due to the contraction of muscles which are attached within the exoskeleton.

[3] The term 'hypodermis' frequently applied to this layer is misleading. The layer is the true outer skin--ectoderm or epidermis.

Now this jointed exoskeleton--an admirably formed suit of armour though it often is--has one drawback: it is not part of the insect's living tissues. It is a cuticle formed by the solidifying of a fluid secreted by the epidermal cells, therefore without life, without the power of growth, and with only a limited capacity for stretching. It follows, therefore, that at least during the period through which the insect continues to grow, the cuticle must be periodically shed. Thus in the life-story of an insect or other arthropod, such as a lobster, a spider, or a centipede, there must be a succession of cuticle-castings-- 'moults' or ecdyses as they are often called.

When such a moult is about to take place the cuticle separates from the underlying epidermis, and a fluid collects beneath. A delicate new cuticle (see fig. 10 _cu'_) is then formed in contact with the epidermis, and the old cuticle opens, usually with a slit lengthwise along the back, to allow the insect in its new coat to emerge. At first this new coat is thin and flabby, but after a period of exposure to the air it hardens and darkens, becoming a worthy and larger successor to that which has been cast. The cuticle moreover is by no means wholly external. The greater part of the digestive canal and the whole air-tube system are formed by inpushings of the outer skin (ectoderm) and are consequently lined with an extension of the chitinous cuticle which is shed and renewed at every moult.

In all insects these successive moults tend to be associated with change of form, sometimes slight, sometimes very great. The new cuticle is rarely an exact reproduction of the old one, it exhibits some new features, which are often indications of the insect's approach towards maturity. Even in some of those interesting and primitive insects the Bristle-tails (Thysanura) and Spring-tails (Collembola), in which wings are never developed, perceptible differences in the form and arrangement of the abdominal limbs can be traced through the successive stages, as R. Heymons (1906) and K.W. Verhoeff (1911) have shown for Machilis. But the changes undergone by such insects are comparatively so slight, that the creatures are often known as 'Ametabola' or insects without transformation in the life-history. Now there are a considerable number of winged insects--cockroaches and grasshoppers for example--in which the observable changes are also comparatively slight. We will sketch briefly the main features of the life-story of such an insect.

The young creature is hatched from the egg in a form closely resembling, on the whole, that of its parent, so that the term 'miniature adult' sometimes applied to it, is not inappropriate. The baby cockroach (fig. 4 _d_) is known by its flattened body, rounded prothorax, and stiff, jointed tail-feelers or cercopods; the baby grasshopper by its strong, elongate hind-legs, adapted, like those of the adult, for vigorous leaping. During the growth of the insect to the adult state there may be four or five moults, each preceded and succeeded by a characteristic instar[4]. The first instar differs, however, from the adult in one conspicuous and noteworthy feature, it possesses no trace of wings. But after the first or the second moult, definite wing-rudiments are visible in the form of outgrowths on the corners of the second and third thoracic segments. In each succeeding instar these rudiments become more prominent, and in the fourth or the fifth stage, they show a branching arrangement of air-tubes, prefiguring the nervures of the adult's wing (fig. 5). After the last moult the wings are exposed, articulated to the segments that bear them, and capable of motion. Having been formed beneath the cuticle of the wing-rudiments of the penultimate instar, the wings are necessarily abbreviated and crumpled. But during the process of hardening of the cuticle, they rapidly increase in size, blood and air being forced through the nervures, so that the wings attaining their full expanse and firmness, become suited for the function of flight.

[4] The convenient term 'instar' has been proposed by Fischer and advocated by Sharp (1895) for the form assumed by an insect during a stage of its life-story. Thus the creature as hatched from the egg is the first instar, after the first moult it has become the second instar, and so on, the number of moults being always one less than the number of instars.

The changes through which these insects pass are therefore largely connected with the development of the wings. It is noteworthy that in an immature cockroach the entire dorsal cuticle is hard and firm. In the adult, however, while the cuticle of the prothorax remains firm, that of the two hinder thoracic and of all the abdominal segments is somewhat thin and delicate on the dorsal aspect. It needs not now to be resistant, because it is covered by the two firm forewings, which shield and protect it, except when the insect is flying. There are, indeed, slight changes in other structures not directly connected with the wings. In a young grasshopper, for example, the feelers are relatively stouter than in the adult, and the prothorax does not show the specifically distinctive shape with its definite keels and furrows. Changes in the secondary sexual characters may also be noticed. For instance, in an immature cockroach both male and female carry a pair of jointed tail-feelers or cercopods on the tenth abdominal segment, and a pair of unjointed limbs or stylets on the ninth. In the adult stage, both sexes possess cercopods, but the males only have stylets, those of the female disappearing at the final moult.

Reviewing the main features of the life-story of a grasshopper or cockroach, we notice that there is no marked or sudden change of form. The newly-hatched insect resembles generally its parent, except that it has no wings. Wing-rudiments appear, however, in an early instar as visible outgrowths on the thoracic segments, and become larger after each moult. All through its various stages the immature insect--nymph as it is called--lives in the same kind of situations and on the same kind of food as its parent, and it is all along active and lively, undergoing no resting period like the pupal stage in the transformation of the butterfly.

One interesting and suggestive fact remains to be mentioned. There are grasshoppers and cockroaches in which the changes are even less than those just sketched, because the wings remain, even in the adult, in a rudimentary state (as for example in the female of the common kitchen cockroach, Blatta orientalis, see fig. 4 _a_), or are never developed at all. Such exceptional winglessness in members of a winged family can only be explained by the recognition of a life-story, not merely in the individual but in the race. We cannot doubt that the ancestors of these wingless insects possessed wings, which in the course of time have been lost by the whole species or by the members of the female sex. It is generally assumed that this loss has been gradual, and so in many cases it probably may have been. But there are species of insects in which some generations are winged and others wingless; a winged mother gives birth to wingless offspring, and a wingless parent to young with well-developed wings. Such discontinuity in the life-story of a single generation forces us to recognise the possibility of similar sudden mutations in the course of that age-long process of evolution to which the facts of insect growth, and indeed of all animal development, bear striking testimony.

CHAPTER III

THE LIFE-STORIES OF SOME SUCKING INSECTS

We may now turn our attention to some examples of the remarkable alternation of winged and wingless generations in the yearly life-cycle of the same species, mentioned at the end of the last chapter. Cockroaches and grasshoppers belong to an order of insects, the Orthoptera[5], characterised by firm forewings and biting jaws; in all of them the change of form during the life-history is comparatively slight. A great contrast to those insects in the structure of the mouth-parts is presented by the Hemiptera, an order including the bugs, pond-skaters, cicads, plant-lice, and scale-insects. These all have an elongated, grooved labium projecting from the head in form of a beak, within which work, to and fro, the slender needle-like mandibles and maxillae by means of which the insect pierces holes through the skin of a leaf or an animal, and is thus enabled to suck a meal of sap or blood, according to its mode of life. In many Hemiptera--the various families of bugs both aquatic and terrestrial, for example--the life-history is nearly as simple as that of a cockroach. It is the family of the plant-lice (Aphidae) that affords typical illustrations of that alternation of generations to which reference has been made.

[5] See outline classification of insects, p. 122.

The yearly cycle of the common Aphids of the apple tree has been lately worked out in detail by J.B. Smith (1900) and E.D. Sanderson (1902). In late autumn tiny wingless males and females are found in large numbers on the withered leaves. The sexes pair together, and the females lay their relatively large, smooth, hard-coated black eggs on the twigs; these resistant eggs carry the species safely over the winter. At springtide, when the leaves begin to sprout from the opening buds the aphid eggs are hatched, and the young insects after a series of moults, through which hardly any change of form is apparent, all grow into wingless 'stem-mothers' much larger than the egg-laying females of the autumn. The stem-mothers have the power, unusual among animals as a whole, but not very infrequent in the insects and their allies, of reproducing their kind without having paired[6] with a male. Eggs capable of parthenogenetic development, produced in large numbers in the ovaries of these females, give rise to young which, developing within the body of the mother, are born in an active state. Successive broods of these wingless virgin females (fig. 6 _a_) appear through the spring and summer months, and as the rate of their development is rapid, often the whole life-story is completed within a week. The aphid population increases very fast. Later a generation appears in which the thoracic segments of the nymphs are seen to bear wing-rudiments like those of the young cockroach, and a host of winged females (fig. 6_b_) are produced; these have the power of migrating to other plants. We understand that wings are not necessary to the earlier broods whose members have plenty of room and food on their native shoots, but that when the population becomes crowded, a winged brood capable of emigration is advantageous to the race.

[6] Such virgin reproduction is termed 'parthenogenesis.'

Many generations of virgin female aphids, some wingless, others winged when adult, succeed each other through the summer months. At the close of the year the latest brood of these bring forth young, which develop into males and egg-laying females; thus the yearly cycle is completed. Variations in points of detail may be noticed in different species of aphids. The autumn males and egg-laying females are, for example, frequently winged, and the same species may have constantly recurring generations of different forms adapted for different food-plants, or for different regions of the same food-plant. But taking a general view of the life-story of aphids for comparison with the life-story of other insects, three points are especially noteworthy. Virgin reproduction recurs regularly, parthenogenetic broods being succeeded by a single sexual brood. A winged parent brings forth young which remain always wingless, and wingless adults produce young which acquire wings. The wings are developed, as in the cockroach, from outward and visible wing-rudiments.

A family of Hemiptera, related to the Aphidae and equally obnoxious to the gardener, is that of the Coccidae or scale-insects. These furnish an excellent illustration of features noticeable in certain insect life-histories. In the first place, the newly-hatched young differs markedly from the parent in the details of its structure. A young coccid (fig. 7 _c_) is flattened oval in shape, has well-developed feelers (fig. 7 _d_) and legs, and runs actively about, usually on the leaves or bark of trees and shrubs, through which it pierces with its long jaws, so that it may suck sap from the soft tissues beneath. After a time it fixes itself by means of these jaws and the characteristic scale or protective covering, composed partly of a waxy secretion and partly of dried excrement, begins to grow over its body. The female loses legs and feelers, and never acquires wings, becoming little more than a sluggish egg-bag (fig. 7 _e_). The male on the other hand passes into a second larval stage in which there are no functional legs, but rudiments of legs and of wings are present on the epidermis beneath the cuticle, as shown by B.O. Schmidt for

Aspidiotus (1885). The penultimate instar of this sex in which the wing-rudiments are visible externally lies passively beneath the scale, its behaviour resembling that of a butterfly pupa. The adult winged male (fig. 7 _a_) leads a short, but active life.

Another family allied to the Aphidae is that of the Cicads, hardly represented in our fauna but abundant in many of the warmer regions of the earth. Here also the young insect differs widely from its parent in form, living underground and being provided with strong fore-legs for digging in the soil. After a long subterranean existence, usually extending over several years, the insect attains the penultimate stage of its life-story, during which it rests passively within an earthen cell, awaiting the final moult, which will usher in its winged and perfect state.

In the life-histories of cicads and coccids, then, there are some features which recall those of the caterpillar's transformation into the butterfly. The newly-hatched insect is externally so unlike its parent that it may be styled a larva. The penultimate instar is quiescent and does not feed. But while the caterpillar shows throughout its life no outward trace of wings, external wing-rudiments are evident in the young stages of the cicad. In the male coccid we find a late larval stage with hidden wing-rudiments, the importance of which, for comparison with the caterpillar, will be appreciated later.

CHAPTER IV

FROM WATER TO AIR

Insects as a whole are preeminently creatures of the land and the air. This is shown not only by the possession of wings by a vast majority of the class, but by the mode of breathing to which reference has already been made (p. 2), a system of branching air-tubes carrying atmospheric air with its combustion-supporting oxygen to all the insect's tissues. The air gains access to these tubes through a number of paired air-holes or spiracles, arranged segmentally in series.

It is of great interest to find that, nevertheless, a number of insects spend much of their time under water. This is true of not a few in the perfect winged state, as for example aquatic beetles and water-bugs ('boatmen' and 'scorpions') which have some way of protecting their spiracles when submerged, and, possessing usually the power of flight, can pass on occasion from pond or stream to upper air. But it is advisable in connection with our present subject to dwell especially on some insects that remain continually under water till they are ready to undergo their final moult and attain the winged state, which they pass entirely in the air. The preparatory instars of such insects are aquatic; the adult instar is aerial. All may-flies, dragon-flies, and caddis-flies, many beetles and two-winged flies, and a few moths thus divide their life-story between the water and the air. For the present we confine attention to the Stone-flies, the May-flies, and the Dragon-flies, three well-known orders of insects respectively called by systematists the Plecoptera, the Ephemeroptera and the Odonata.

In the case of many insects that have aquatic larvae, the latter are provided with some arrangement for enabling them to reach atmospheric air through the surface-film of the water. But the larva of a stone-fly, a dragon-fly, or a may-fly is adapted more completely than these for aquatic life; it can, by means of gills of some kind, breathe the air dissolved in water.

The aquatic young of a stone-fly does not differ sufficiently in form from its parent to warrant us in calling it a larva; the life-history is like that of a cockroach, all the instars however except the final one--the winged adult or _imago_--live in the water. The young of one of our large species, a Perla for example, has well-chitinised cuticle, broad head, powerful legs, long feelers and cerci like those of the imago; its wings arise from external rudiments, which are conspicuous in the later aquatic stages. But it lives completely submerged, usually clinging or walking beneath the stones that lie in the bed of a clear stream, and examination of the ventral aspect of the thorax reveals six pairs of tufted gills, by means of which it is able to breathe the air dissolved in the water wherein it lives. At the base of the tail-feelers or cerci also, there are little tufts of thread-like gills as J.A. Palm 閣 (1877) has shown. An insect that is continually submerged and has no contact with the upper air cannot breathe through a series of paired spiracles, and during the aquatic life-period of the stone-fly these remain closed. Nevertheless, breathing is carried on by means of the ordinary system of branching air-tubes, the trunks of which are in connection with the tufted hollow gill-filaments, through whose delicate cuticle gaseous exchange can take place, though the method of this exchange is as yet very imperfectly understood. When the stone-fly nymph is fully grown, it comes out of the water and climbs to some convenient eminence. The cuticle splits open along the back, and the imago, clothed in its new cuticle, as yet soft and flexible, creeps out. The spiracles are now open, and the stone-fly breathes atmospheric air like other flying insects. But throughout its winged life, the stone-fly bears memorials of its aquatic past in the little withered vestiges of gills that can still be distinguished beneath the thorax.

The adult dragon-fly (fig. 8 _d_) is specialised in such a way that it captures its prey--flies and other small insects--on the wing, swooping through the air like a hawk and feeding voraciously. The head is remarkable for its large globular compound eyes, its short bristle-like feelers, and its very strong mandibles which bite up the bodies of the victims. The thorax bears the two pairs of ample wings, firm and almost glassy in texture, and its segments are projected forward ventrally, so that all six legs, which are armed with rows of sharp, slender spines, can be held in front of the mouth, where they form an effective fly-trap. The abdomen is very long and usually narrow.

A female dragon-fly after a remarkable mode of pairing, the details of which are beside our present subject, drops her eggs in the water, or lays them on water-weeds, perhaps cutting an incision where they can be the more safely lodged, or even goes down below the surface and deposits them in the mud at the bottom of a pond. From the eggs are hatched the aquatic larvae which differ in many respects from the imago. The dragon-fly larva has the same predaceous mode of life as its parent, but it is sluggish in habit, lurking for its prey at the bottom of the pond, among the mud or vegetation, which it resembles in colour. The thoracic segments have not the specialisation that they show in the imago; the abdomen is relatively shorter and broader. The larval head has, like that of the imago, short feelers, and the eyes are somewhat large, though far from attaining the size of the great globular eyes of the dragon-fly. But the third pair of jaws, forming the labium, are most remarkably modified into a 'mask,' the distal central portion (mentum) being hinged to the basal piece (sub-mentum) which is itself jointed below the head. The mentum carries at its extremity a pair of lobes with sharp fangs. Thus the mask can be folded under the head when the larva lurks in its hiding place, or be suddenly darted out so as to secure any unwary small insect that may pass close enough for capture. Dragon-fly larvae walk, and also swim by movements of the abdomen or by expelling a jet of water from the hind-gut.

The walls of this terminal region of the intestine have areas lined with delicate cuticle and traversed by numerous air-tubes, so that gaseous exchange can take place between the air in the tubes and that dissolved in the water. The larvae of the larger and heavier dragon-flies (Libellulidae and Aeschnidae) breathe mostly in this way. Those of the slender and delicate 'Demoiselles' (Agrionidae) are provided with three leaf-like gill-plates at the tail, between whose delicate surfaces numerous air-tubes ramify. These gill-plates are at times used for propulsion. Thus air supply is ensured during aquatic life. But occasionally, when the water in which the larva lives is foul and poor in oxygen, the tail is thrust out of the water so that air can be admitted directly into the intestinal chamber. The aquatic life of these insects lasts for more than a year, and F. Balfour-Browne (1909) has observed from ten to fourteen moults in Agrion. Outward wing-rudiments are early visible on the thoracic segments; when these have become conspicuous the insect, beginning in some respects to approach the adult condition, is often called a nymph. In an advanced dragon-fly nymph, H. Dewitz (1891) has shown that the thoracic spiracles are open, and, as the time for its final moult draws near, the insect may thrust the front part of its body out of the water, and breathe atmospheric air through these. Thus before the great change takes place the nymph has foretastes of the aerial mode of breathing which it will practise when the perfect stage shall have been attained. The emergence of the dragon-fly from its nymph-cuticle has been described by many naturalists from de Rumur (1740) to L.C. Miall (1895) and O.H. Latter (1904). The nymph climbs out of the water by ascending some aquatic plant, and awaits the change so graphically sketched by Tennyson:

A hidden impulse rent the veil, Of his old husk, from head to tail, Came out clear plates of sapphire mail.

'From head to tail,' for the nymph-cuticle splits lengthwise down the back, and the head and thorax of the imago are freed from it (fig. 8 _a_), then the legs clasp the empty cuticle, and the abdomen is drawn out (fig. 8 _b, c_). After a short rest, the newly-emerged fly climbs yet higher up the water-weed, and remains for some hours with the abdomen bent concave dorsalwards (fig. 8 _d_), to allow space for the expansion and hardening of the wings. For some days after emergence the cuticle of the dragon-fly has a dull pale hue, as compared with the dark or brightly metallic aspect that characterises it when fully mature. The life of the imago endures but a short time compared with the long aquatic larval and nymphal stages. After some weeks, or at most a few months, the dragon-flies, having paired and laid their eggs, die before the approach of winter.

The life-story of a may-fly follows the same general course as that just described for the dragon-flies, but there are some suggestive differences. In the first place, we notice a wider divergence between the imago and the larva. An adult may-fly is one of the most delicate of insects; the head has elaborate compound eyes, but the feelers are very short, and the jaws are reduced to such tiny vestiges that the insect is unable to feed. Its aquatic larva is fairly robust, with a large head which is provided with well-developed jaws, as the larval and nymphal stages extend over one or two years, and the insects browse on water-weeds or devour creatures smaller and weaker than themselves. They breathe dissolved air by means of thread-like or plate-like gills traversed by branching air-tubes, somewhat resembling those of the demoiselle dragon-fly larva. But in the may-fly larva, there is a series of these gills (fig. 9_b_) arranged laterally in pairs on the abdominal segments, and C. Boner (1909) has recently given reasons, from the position and muscular attachments of these organs, for believing that they show a true correspondence to (in technical phraseology are homologous with) the thoracic legs. One feature in which the larva often agrees with the imago is the possession on the terminal abdominal segment of a pair of long jointed cerci, and in many genera a median jointed tail-process (see fig. 9) is also present, in some cases both in the larva and the imago, in others in the larva during its later stages only. The prolonged larval life in may-flies often involves a large series of moults; Lubbock (1863) has enumerated twenty-one in the life-history of Chloeon. In the second year of aquatic life wing-rudiments (fig. 9 _a_) are visible, and the larva becomes a nymph. When the time for the winged condition approaches the nymphs leave the water in large swarms. The vivid accounts of these swarms given by Swammerdam (1675), de Rumur (1742) and other old-time observers are available in summarised form for English readers in Miall's admirable book (1895). May-flies are eagerly sought as food by trout, and the rise of the fly on many lakes ushers in a welcome season to the angler.

The nymph-cuticle opens and the winged insect emerges. But this is not the final instar; may-flies are exceptional among insects in undergoing yet another moult after they have acquired wings which they can use for flight.

The instar that emerges from the nymph-cuticle is a sub-imago, dull in hue, with a curious immature aspect about it. A few hours later the final moult takes place, a very delicate cuticle being shed and revealing the true imago. Then follow the dancing flight over the calm waters, the mating and egg-laying, the rapid death. The whole winged existence prepared for by the long aquatic life may be over in a single evening; at most it lasts but for a few days.

[Illustration: Fig. 9. Nymph of May-fly (_Chloeon dipterum_) showing on right side wing-rudiment (_a_), on left tracheal gills (_b_). Magnified 4 times. [Feelers and legs are cut short.] From Miall and Denny after Vayssie.]

In the development of the may-flies, then, we notice not only a considerable divergence between larva and imago, both in habitat and structure; we see also what is to be observed often in more highly organised insects--a feeding stage prolonged through the years of larval and nymphal life, while the winged imago takes no food and devotes its energies through its short existence to the task of reproduction. Such division of the life-history into a long feeding, and a short breeding period has, as will be seen later, an important bearing on the question of insect transformation generally, and the dragon-flies and may-flies afford examples of two stages in its specialisation. The sub-imaginal instar of the may-fly furnishes also a noteworthy fact for comparison with other insect histories. In two points, however, the life-story of these flies with their aquatic larvae recalls that of the cockroach. All the larval and nymphal instars are active, and the wing-rudiments are outwardly visible long before the final moult.

CHAPTER V

TRANSFORMATIONS,--OUTWARD AND INWARD

We are now in a position to study in some detail the transformation of those insects whose life-story corresponds more or less closely with that of the butterfly, sketched in the opening pages of this little book. In the case of some of the insects reviewed in the last three chapters, the may-flies and cicads for example, a marked difference between the larva and the imago has been noticed; in others, as the coccids, we find a resting instar before the winged condition is assumed, suggesting the pupal stage in the butterfly's life-story.

The various insect orders whose members exhibit no marked divergence between larva and imago (the Orthoptera for example) are often said to undergo no transformation, to be 'Ametabola.' Those with life-stories such as the dragon-flies' are said to undergo partial transformation, and are termed 'Hemimetabola.' Moths, caddis-flies, beetles, two-winged flies, saw-flies, ants, wasps, bees, and the great majority of insects, having the same type of life-story as the butterfly, are said to undergo complete transformation and are classed as 'Metabola' or 'Holometabola.' Wherein lies the fundamental difference between these Holometabola on the one hand and the Hemimetabola and Ametabola on the other? It is not that the larva differs from the imago or that there is a passive stage in the life-history; these conditions are observable among insects with a 'partial' transformation as we have seen, though the resting instar that simulates the butterfly pupa is certainly exceptional. It has been pointed out by Sharp (1899) that the most important indication of the difference between the two modes of development is furnished by the position of the wing-rudiments. In all Ametabola and Hemimetabola these are visible externally long before the penultimate instar has been reached; in the Holometabola they are not seen until the pupal stage.

Attention has already been drawn to the contrast in outward form between a butterfly and its caterpillar. As in the case of dragon-fly or may-fly, the larval period is essentially a time for feeding and growth, and during this period the larval cuticle is cast four or five, in some species even seven or eight times. After each moult some changes in detail may be observable, for example in the proportions of the body-segments or their outgrowths, in the colour or the closeness of the hairy or spiny armature. But in all main features the caterpillar retains throughout its life the characteristic form in which it left the egg. From the tiny, newly-hatched larva to the full-fed caterpillar, possibly several inches in length, there is all along the same crawling, somewhat worm-like body, destitute of any outward trace of wings. When however the last larval cuticle has split open lengthwise along the back, and has been worked off by vigorous wriggling motions of the insect, the pupa thus revealed shows the wing-rudiments conspicuous at the sides of the body, and lying neatly alongside these are to be seen the forms of feelers, legs, and maxillae of the imago prefigured in the cuticle of the pupa (fig. 1 _e_). The pupa thus resembles the imago much more closely than it resembles the larva; even in the proportions of the body a relative shortening is to be noticed, and the imago of any insect with complete transformation is reduced in length as compared with the full-fed larva. Now these wings and other structures characteristic of the imago, appear in the pupa which is revealed by the shedding of the last larval cuticle. From these facts we infer that the wing-rudiments must be present in the larva, hidden beneath the cuticle; and until the last larval instar, not beneath the cuticle only, but growing in such-wise that they are hidden by the epidermis. For if they were growing outwardly the new cuticle would be formed over them, so that they would be apparent after the next moult. But it is clear that only in the pupa, forming beneath the cuticle of the last larval instar, can they grow outwards.

Anatomical study of the caterpillar at various stages verifies the conclusions just drawn from superficial observation. A hundred and fifty years ago P. Lyonet in his monumental work (1762) on the caterpillar of the Goat Moth (Cossus) detected, in the second and third thoracic segments, four little white masses buried in the fat-body, and, while doubtful as to their real meaning, he suggested that their number and position might well give rise to the suspicion that they were rudiments of the wings of the moth. But it was a century later that A. Weismann in his classical studies (1864) on the development of common flies, showed the presence in the maggot of definite rudiments of wings, and other organs of the adult--rudiments to which he gave the name of imaginal discs. We will recur later to these transformations of the Diptera. For the present, we pursue our survey of changes in the life-history of the Lepidoptera and can take to guide us the excellent researches of J. Gonin (1894).

Careful study of the imaginal discs of the wings in a caterpillar (fig. 10) made by examining microscopically sections cut through them, shows that the epidermis is pushed in to form a little pouch (_C, p_) and that into this grows the actual wing-rudiment. Consequently the whitish disk which seems to lie within the body-wall of the larva, is really a double fold of the epidermis, the outer fold forming the pouch, the inner the actual wing-bud. Into the cavity of the latter pass branches from the air-tube system. In its earliest stage, the wing-bud is simply an ingrowing mass of cells (fig. 10 _A_) which subsequently becomes an inpushed pouch (_B_). Until the last stage of larval life the wing-bud remains hidden in its pouch, and no cuticle is formed over it. When the pupal stage draws near the bud grows out of its sheath, and projecting from the general surface of the epidermis becomes covered with cuticle to be revealed, as we have seen, after the last larval moult, as the pupal wing. Thus all through the life of the humble, crawling caterpillar, 'it doth not yet appear what it shall be,' but there are being prepared, hidden and unseen, the wondrous organs of flight, which in due time will equip the insect for the glorious aerial existence that awaits it.

As mentioned above, this hidden growth of the wing-rudiments, in butterflies, beetles, flies, bees, and the great majority of the winged insects, has been emphasised by Sharp (1899) as a character contrasting markedly with the outward and visible growth of the wing-rudiments in such insects as cockroaches, bugs, and dragon-flies. The divergence between the two modes of development is certainly very striking, and a conceivable method of transition from the one to the other is not easy to explain. Sharp has expressed the divergence by the terms Endopterygota, applied to all the orders of insects with hidden wing-rudiments (the 'Metabola' or 'Holometabola' of most classifications) and Exopterygota, including all those insects whose wing-rudiments are visible throughout growth ('Hemimetabola' and 'Ametabola'). Those curious lowly insects, belonging to the two orders of the Collembola and Thysanura, none of whose members ever develop wings at all, form a third sub-class, the Apterygota (see Classificatory Table, p. 122).

Not the wings only, but other structures of the imago, varying in extent in different orders, are formed from the imaginal discs. For example, de Rumur and G. Newport (1839) found that if the thoracic leg of a late-stage caterpillar were cut off, the corresponding leg of the resulting butterfly would still be developed, although in a truncated condition. Gonin has shown that in the Cabbage White butterfly (_Pieris brassicae_) the legs of the imago are represented, through the greater part of larval life, only by small groups of cells situated within the bases of the larval legs. After the third moult these imaginal discs grow rapidly and the proximal portion of each, destined to develop into the thigh and shin of the butterfly's leg, sinks into a depression at the side of the thorax, while the tip of the shin and the five-segmented foot project into the cavity of the larval leg. Hence we understand that the amputation of the latter by the old naturalists truncated only and did not destroy the imaginal limb. In the blow-fly maggot, Weismann, B.T. Lowne (1890) and J. Van Rees (1888) have shown that the imaginal discs of the legs (fig. 11--1, 2, 3) grow out from deep dermal inpushings. Simple at first, these outgrowths by partial splitting, become differentiated into thigh and shin.

Similarly the feelers and jaws of the butterfly are developed from imaginal discs, and this fact explains how it comes to pass that they differ so widely from the corresponding structures in the caterpillar. The larval feelers (fig. 3 _At_) are short and stumpy, those of the butterfly long and many-jointed. The maxilla of the larva (fig. 3 _Mx_) consists of a base carrying two short jointed processes; in the butterfly a certain portion of the maxilla, the hood or galea, is modified into a long, flexible grooved process, capable of forming with its fellow the trunk through which the insect sucks its liquid food (fig. 2). Nothing but some such provision as that of the imaginal discs could render possible the wonderful replacement of the caterpillar's jaws, biting solid food, into those of the butterfly sipping nectar from flowers.

A curious segmental displacement of the imaginal discs with regard to the larva is noticeable in some Diptera. In the larva of the harlequin-midge (Chironomus) as described by Miall and Hammond (1900) the brain is situated in the thorax, and the imaginal discs for the head, eyes, and feelers of the adult lie in close association with it, though they arise from inpushings of the larval head. These rudiments do not appear until the last larval stage has been reached. In the gnats Culex and Corethra, on the other hand, the imaginal discs for the head-appendages retain their normal position within the larval head, and appear in an early stage of larval life. Among the flies of the bluebottle group (Muscidae) the brain (fig. 11 _B_) is situated, as in Chironomus, in the thoracic region of the legless maggot, which is the larva of an insect of this family, and the imaginal discs for eyes and feelers (fig. 11 e, _f_) lie just in front of it. Here, the imaginal buds of the legs (fig. 11--1, 2, 3) and wings (fig. 11 W, _w_) are deeply inpushed, retaining their connection with the skin only by means of a thread of cells. As the larva is legless and headless its outer form is not affected by the discs and it is not surprising to learn that they appear early. It has indeed been suggested that the pharyngeal region of the larva, in connection with which the imaginal head-discs are developed, should be regarded, though it lies in the thorax, as an inpushed anterior section of the larval head. In any case this region is pushed out during the formation of the pupa within the final larval cuticle, so that the imaginal head with its contained brain, its compound eyes, and its complex feelers, takes its rightful place at the front end of the insect.

The mention of the brain suggests a few brief remarks on the changes in the internal organs during insect transformation. There are no imaginal discs for the nervous system; the brain, nerve-cords and ganglia of the butterfly or bluebottle are the direct outcome of those of the caterpillar or maggot. More than seventy years ago, Newport (1839) traced the rapid but continuous changes, which, during the early pupal period, convert the elongate nerve-cord of the caterpillar with its relatively far-separated ganglia into the shortened, condensed nerve-cord of the Tortoise-shell butterfly (_Vanessa urticae_) with several of the ganglia coalesced. In many Diptera, on the other hand, the nervous system of the larva is more concentrated than that of the imago.

The tubular heart also of a winged insect is the directly modified survival of the larval heart.

Similarly the reproductive organs undergo a gradual, continuous development throughout an insect's life-story. Their rudiments appear in the embryo, often at a very early stage; they are recognisable in the larva, and the matured structures in the imago are the result of their slow process of growth, the details of which must be reckoned beyond the scope of this book. For a full summary of the subject the reader is referred to L.F. Henneguy's work (1904) containing references to much important modern literature, which cannot be mentioned here.

On the other hand, the digestive system of insects that undergo a metamorphosis, passes through a profound crisis of dissolution and rebuilding. This is not surprising when we remember that there is often a great difference between larva and imago in the nature of the food. The digestive canal of a caterpillar runs a fairly straight course through the body and consists of a gullet, stomach (mid-gut), intestine, and rectum; it is adapted for the digestion of solid food. In the butterfly there is one outgrowth of the gullet in the head--a pharyngeal sac adapted for sucking liquids; and another outgrowth at the hinder end of the gullet (which is much longer than in the larva)--a crop or food-reservoir lying in the abdomen. The intestine of the butterfly also is longer than that of the larva, being coiled or twisted. Towards the end of the last larval stage, the cells of the inner coat (epithelium) lining the stomach begin to undergo degeneration, small replacing cells appearing between their bases and later giving rise to the more delicate epithelium that lines the mid-gut of the imago. The larval cells are shed into the cavity of the stomach and become completely broken down. J. Anglas (1902), describing these microscopic changes in the transformations of wasps and bees, has shown that the tiny replacing cells can be recognised in sections through the digestive canal of a very young larva; they may be regarded as representing imaginal buds of the adult gastric epithelium. In the transformations of two-winged flies of the bluebottle group, A. Kowalevsky (1887) has shown that these replacing cells are aggregated in little masses scattered at different points along the stomach and thus corresponding rather closely to the imaginal discs of the legs and wings.

The gullet, crop, and gizzard of an insect, which lie in front of the stomach, are lined by cells derived from the outer skin (ectoderm) which is pushed in to form what is called the 'fore-gut.' Similarly the intestine and rectum, behind the stomach, are lined with ectodermal cells which arise from the inpushed 'hind-gut.' The larval fore- and hind-guts are broken down at the end of larval life and their lining is replaced by fresh tissue derived from two imaginal bands which surround the cavity of the digestive tube, one at the hinder end of the fore-gut, and the other at the front end of the hind-gut. The larval salivary glands in connection with the gullet are also broken down, and fresh glands are formed for the imago.

A large part of the substance of an insect larva consists of muscular tissue, surrounding the digestive tube, and forming the great muscles that move the various parts of the body, and of fat, surrounding the organs and serving as a store of food-material. Very many of the muscle-fibres and the fat-cells also become disintegrated during the late larval and pupal stages, and the corresponding tissues of the adult are new formations derived from special groups of imaginal cells, though some muscles may persist from the larva to the adult. Similarly the complex air-tube or tracheal system of the larva is broken down and a fresh set of tubes is developed, adapted to the altered body-form of pupa and imago.

The destruction of larval tissue and the development of replacing organs from special groups of cells, derived of course from the embryo, and carrying on the continuity of cell-lineage to the adult, are among the most remarkable facts connected with the life-story of insects. The process of tissue-destruction is known as 'histolysis'; the rebuilding process is called 'histogenesis.' Considerable difference of opinion has existed as to factors causing histolysis, and for a summary of the conflicting or complementary theories, the reader is referred to the work of L.F. Henneguy (1904, pp. 677-684). In the histolysis of the two-winged flies, wandering amoeboid cells--like the white corpuscles or leucocytes of vertebrate blood--have been observed destroying the larval tissues that need to be broken down, as they destroy invading micro-organisms in the body. But students of the internal changes that accompany transformation in insects of other orders have often been unable to observe such devouring activity of these 'phagocytes,' and attribute the dissolution of the larval tissues to internal chemical changes. The fact that in all insect transformation a part, and in many a large part, of the larval organs pass over to the pupa and imago, suggests that only those structures whose work is done are broken down through the action of internally formed destructive substances, and one function of the phagocytes is to act as scavengers by devouring what has become effete and useless.

CHAPTER VI

LARVAE AND THEIR ADAPTATIONS

Among the insects that undergo a complete transformation, there is, as we have seen in the preceding chapter, an amount of inward change, of dissolution and rebuilding of tissues, that varies in its completeness in members of different orders. It is now advisable to consider the various outward forms assumed by the larvae of these insects, or rather by a few examples chosen from a vast array of well-nigh 'infinite variety.'

In comparing the transformations of endopterygote insects of different orders, it is worthy of notice that in some cases all the members of an order have larvae remarkably constant in their main structural features, while in others there is great variety of larval form within the order. For example, the caterpillars of all Lepidoptera are fundamentally much alike, while the grubs of beetles of different families diverge widely from one another. A review of a selected series of beetle-larvae will therefore serve well to introduce this branch of the subject.

Beetles are as a rule remarkable among insects for the firm consistency of their chitinous cuticle, the various pieces (_sclerites_) of which are fitted together with admirable precision. In some families of beetles the larva also is furnished with a complete chitinous armour, the sclerites, both dorsal and ventral, of the successive body-segments being hard and firm, while the relatively long legs possess well-defined segments and are often spiny. Such a larva is evidently far less unlike its parent beetle than a caterpillar is unlike a butterfly. Perhaps of all beetle larvae, the woodlouse-like grub (fig. 12 _b_) of a carrion-beetle (Silpha) or of a semi-aquatic dascillid such as Helodes shows the least amount of difference from the typical adult, on account of the conspicuous jointed feelers. The larval glow-worm, however, is of the same woodlouse-like aspect, and in this case, where the female never acquires wings, but becomes mature in a form which does not differ markedly from that of the larva, the exceptional resemblance is closer still. In all beetle-grubs the legs are simplified, there being only one segment (a combined shin and foot) below the knee-joint, whereas in the adult there is a shin followed by five, four, or at least three distinct tarsal segments. The foot of an adult beetle bears two claws at its tip, while the larval foot in the great majority of families has only one claw. In one section of the order, however, the Adephaga comprising the predaceous terrestrial and aquatic beetles, the larval foot has, like that of the adult, two claws. Some adephagous larvae, notably those of the large carnivorous water-beetles (Dyticus), often destructive to tadpoles and young fish, have completely armoured bodies as well as long jointed legs. More commonly, as with most of the well-known Ground-beetles (Carabidae), the cuticle is less consistently hard, firm sclerites segmentally arranged alternating with considerable tracts of cuticle which remain feebly chitinised and flexible. Most of the adephagous larvae (fig. 13) have a pair of stiff processes on the ninth abdominal segment, and the insect, from its general likeness to a bristle-tail of the genus Campodea, is often called a campodeiform larva (Brauer, 1869). From such as these, a series of forms can be traced among larvae of beetles, showing an increasing divergence from the imago. The well-known wireworms--grubs of the Click-beetles (Elateridae)--that eat the roots of farm crops, have well-armoured bodies, but their shape is elongate, cylindrical, worm-like; and their legs are relatively short, the build of the insect being adapted for rapid motion through the soil. The grubs of the Chafers (Scarabaeidae) are also root-eaters, but they are less active in their habits than the wireworms, and the cuticle of their somewhat stout bodies is, for the most part, pale and flexible; only the head and legs are hard and horny. Usually an evident correspondence can be traced between the outward form of any larva and its mode of life. For example, in the family of the Leaf-beetles (Chrysomelidae) some larvae feed openly on the foliage of trees or herbs, while others burrow into the plant tissues. The exposed larvae of the Willow-beetles (Phyllodecta, fig. 14) have their somewhat abbreviated body segments protected by numerous spine-bearing, firm tubercles. But the grub of the 'Turnip Fly' (Phyllotreta) which feeds between the upper and lower skins of a leaf, or of Psylliodes chrysocephala (fig. 15), which burrows in stalks, has a pale, soft cuticle like that of a caterpillar.

In the larvae of the little timber-beetles and their allies (Ptinidae), including the 'death-watches' whose tapping in old furniture is often heard, a marked shortening of the legs and reduction in the size of the head accompany the whitening and softening of the cuticle. This shortening of the legs is still more marked in the larvae of the Longhorn Beetles (Cerambycidae) burrowing in the wood of trees or felled trunks; here the legs are reduced to small vestiges.

Finally in the large family of the Weevils (Curculionidae, fig. 16) and the Bark-beetles (Scolytidae), the grubs, eating underground root or stem structures, mining in leaves or seeds, or tunnelling beneath the bark of trees, have no legs at all, the place of these limbs being indicated only by tiny tubercles on the thoracic segments. Such larvae as these latter are examples of the type called eruciform by A.S. Packard (1898) who as well as other writers has laid stress on the series of transitional steps from the campodeiform to the eruciform type afforded by the larvae of the Coleoptera.

A fact of much importance in the transformations of beetles as pointed out by Brauer (1869) is that in a few families, the first larval instar is campodeiform, while the subsequent instars are eruciform. We may take as an example of such 'hypermetamorphosis' the life-story of the Oil or Blister-beetles (Meloidae) as first described by J.H. Fabre (1857), and later with more elaboration by H. Beauregard (1890). From the egg of one of these beetles is hatched a minute armoured larva, with long feelers, legs, and cerci, whose task is, for example, to seize hold of a bee in order that the latter may carry it, an uninvited guest, to her nest. Safely within the nest, the little 'triungulin' beetle-grub moults; the second instar has a soft cuticle and relatively shorter legs, which, as the larva, now living as a cuckoo-parasite, proceeds to gorge itself with honey, soon appear still further abbreviated. Later comes a stage during which legs are entirely wanting, the larva then resting and taking no food. The last larval instar again has short legs like the grub of the second period. In connection with this life-history we notice that the newly-hatched larva is not in the neighbourhood of its appropriate food. Hence the preliminary armoured and active instar is necessary in order to reach the feeding place; this journey accomplished, the eruciform condition is at once assumed.

In all cases indeed we may say that the particular larval form is adapted to the special conditions of life. A few examples from other orders of endopterygote insects will illustrate this point. The campodeiform type is relatively unusual, but most of the Neuroptera have larvae of this kind, active, armoured creatures with long legs, though devoid of the tail-processes often associated with similar larvae among the Coleoptera. Such are the 'Ant-lions,' larvae of the exotic lacewing flies, which hunt small insects, digging a sandy pit for their unwary steps in the case of the best-known members of the group, some of which are found as far north as Paris. In our own islands the 'Aphis-lions,' larvae of Hemerobius and Chrysopa, prowl on plants infested with 'green-fly' which they impale on their sharp grooved mandibles, sucking out the victims' juices, and then, in some cases, using the dried cuticle to furnish a clothing for their own bodies. Among these insects, while the mouth of the imago is of the normal mandibulate type adapted for eating solid food, the larval mouth is constricted and the slender mandibles are grooved for the transmission of liquid food.

Turning to eruciform types of larva, we find the caterpillar (fig. 1 b, c, _d_) distinguished by its elongate, usually cylindrical body with feeble cuticle, short thoracic legs and a variable number of pairs of abdominal pro-legs, universal among the moths and butterflies forming the great order Lepidoptera, and usual among the saw-flies, which belong to the Hymenoptera. The vast majority of caterpillars feed on the leaves of plants and their long worm-like bodies with the series of paired pro-legs, are excellently adapted for their habit of clinging to twigs, and crawling along shoots or the edges of leaves as they go in search of food. Of great importance to a caterpillar is its power of spinning silk, consisting of fine threads solidified from the secretion of specially modified salivary glands whose ducts open in the insect's mouth at the tip of the tubular tongue which forms a spinneret.

On the same bush caterpillars of moths and of saw-flies may often be seen feeding together. The lepidopterous caterpillar, in our countries at least, has never more than five pairs of pro-legs, situated on the third, fourth, fifth, sixth, and tenth abdominal segments; each of these pro-legs bears a number of minute hooklets, arranged in a circular or crescentic pattern, which assist the caterpillar in clinging to its food-plant. The saw-fly caterpillar, on the other hand, may have as many as eight pairs of pro-legs, the series beginning on the second abdominal segment; here, however, the pro-legs have no hooklets. Among the Lepidoptera, we notice a reduction in the number of pro-legs in the 'looper' caterpillars of Geometrid moths. Here only two pairs are present, those on the sixth and tenth abdominal segments. Consequently, as the caterpillar can cling only by the thorax and by the hinder region of the abdomen, the middle region of the body is first straightened out and then bent into an arch-like form, as the insect makes its progress by alternate movements of stretching and 'looping.'

Caterpillars, with their relatively soft bodies, feeding openly on the leaves of plants, are exposed to the attacks of many enemies, and the various ways in which they obtain protection are well worth studying. A clothing of hairs[7] or spines is often present, and it is interesting to find that many species of our native Tiger and Eggar Moths (Arctiadae and Lasiocampidae) which pass the winter in the larval stage, have caterpillars with an especially dense hairy covering (fig. 17). Experiments have shown that hairy and spiny insects are distasteful to birds and other creatures that prey readily on smooth-skinned species, a conclusion that might well have been expected. Certain smooth caterpillars however appear to be protected by producing some nauseous secretion, which renders them unpalatable. Many of these, as the familiar cream yellow and black larva of the Magpie Moth (_Abraxas grossulariata_), are very conspicuously adorned, and furnish examples of what is known as 'warning coloration,' on the supposition that the gaudy aspect of such insects serves as an advertisement that they are not fit to eat, and that birds and other possible devourers thus learn to leave them alone. On the other hand, smooth caterpillars which are readily eaten by birds are usually 'protectively' coloured, so as to resemble their surroundings and remain hidden except to careful seekers. Many such caterpillars are green, the upper surface, which is naturally exposed to the light, being darker than the lower which is in shadow. When the caterpillar is large, the green area is often broken up by pale lines, longitudinal as on the larvae of many Owl Moths (Noctuidae) or oblique, as on the great caterpillars of most Hawk Moths (Sphingidae). Such an arrangement tends to make the insect less easily seen than were it to display a continuous area of the same colour. The 'looper' caterpillars mentioned above afford remarkable examples of 'protective' resemblance, for many of them show a marvellous likeness to the twigs of their food-plant, tubercles on the insect's body resembling closely the little outgrowths of the plant's cortex. It has been shown by E.B. Poulton (1892) that many caterpillars are, in their early stages, directly responsive to their surroundings as regards colour. Usually green when hatched, they remain green if kept among leaves or young shoots of plants, while they turn red, brown, or blackish if placed among twigs of these respective hues. This effect appears to be due to a direct response of the subcutaneous tissue to the rays of light reflected from the surrounding objects. The sensitiveness dies away as the caterpillar grows older, since little or no change of hue in response to a change of environment could be induced after the penultimate moult.

[7] The 'hairs' of an insect are not in the least comparable to the hairs of mammals, being in truth, modified portions of the cuticle, secreted by special cells.

Among those families of the Lepidoptera which are usually regarded as low in the scale of organisation, caterpillars are very generally protected by the habit of feeding in some concealed situation. For example, the great larvae of the Goat Moth (Cossus) and the whitish caterpillars of the Clearwing Moths (Sesiidae) burrow through the wood of trees, eating the timber as they go. The little irritable caterpillars of the Bell Moths (Tortricidae) roll leaves, fastening the edges together with silk, and thus make for themselves a shelter; or they bore their way into seeds or fruits, like the larva of the Codling Moth that is the cause of 'worm-eaten' apples, too well-known to orchard-keepers. Very many small caterpillars mine between the two skins of a leaf, eating out the soft green tissue, and giving rise to a characteristic blister in form of a spreading patch or a narrow sinuous track through the leaf. The caterpillars of the Clothes-moths (Tineidae) make for themselves garments out of their own excrement, the particles fastened together by silk. In such curious cylindrical cases they wander over the wool or fur, feeding and indirectly supplying themselves with clothing at the same time.

The case-forming habit of the Clothes-moth caterpillars leads us naturally to consider the similar habit adopted by their allies the Caddis-larvae which live in the waters of ponds and streams, for the Caddis-flies (Trichoptera) have much in common with the more primitive Lepidoptera. The caddis-larva is as a rule of the eruciform type, but with well-developed thoracic legs, and with hook-like tail-appendages; by means of the latter it anchors itself to the extremity of its curious 'house.' It is of interest to note that in the earlier stages of some caddises lately described and figured by A.J. Siltala (1907), the legs are relatively very long, and the larva is quite campodeiform in aspect. Some of these caddis-grubs retain the campodeiform condition and do not shelter permanently in cases, as their relations do. Different genera of caddises differ in their mode of building. Some fasten together fragments of water-weeds and plant refuse, others take tiny particles of stone, of which they make firmly compacted walls, others again lay hold of water-snail shells, which may even contain live inhabitants, and bind these into a limy rampart behind which their bodies are in safe hiding.

The silk with which the 'caddis-worms' fasten together the materials for their houses is produced from spinning-glands which like those of the Lepidoptera open into the mouth.

The survey of the various types of beetle-larvae enumerated above (pp. 50-56) concluded with a short description of the legless grub, which is the young form of a weevil or a bark-beetle. This is a larva in which the head alone has its cuticle firm and hard; the rest of the body is covered with a pale, flexible cuticle, so that the grub is often described as 'fleshy.' This type of larva is by no means confined to certain families of the beetles, it is frequently met with, in more or less modified form, in two other important orders of insects, the Hymenoptera and the Diptera. Among the Hymenoptera this is indeed the predominant larval type. We have just seen that a caterpillar is the usual form of larva among the saw-flies, but in all other families of the Hymenoptera we find the legless grub. A grub of this order may usually be distinguished from the larva of a weevil or other beetle, by its relatively smaller head and smoother, less wrinkled cuticle; it strikes the observer as a feebler, more helpless creature than a beetle-grub. And it is of interest to note that this somewhat degraded type of larva is remarkably constant through a great series of families--gall-flies, ichneumon-flies, wasps, bees (fig. 18), ants--that vary widely in the details of their structure and in their habits and mode of life. Almost without exception, however, they make in some way abundant provision for their young. The feeble, helpless, larva is in every case well sheltered and well fed; it has not to make its own way in the world, as the active armoured larva of a ground-beetle or the caterpillar of a butterfly is obliged to do.

Among those saw-flies whose larvae feed throughout life in a concealed situation, we find an interesting transition between the caterpillar and the legless grub. For example, the giant saw-flies (so called 'Wood-wasps') have larvae that burrow in timber, and these larvae possess relatively large heads, somewhat flattened bodies with pointed tail-end, and very greatly reduced legs. The feeble legless grub, characteristic of the remaining families of the Hymenoptera, is provided for in a well-nigh endless variety of ways. The female imago among these insects is furnished with an elaborate and beautifully formed ovipositor, and the act of egg-laying is usually in itself a provision for the offspring. Gall-flies pierce plant-tissues within which their grubs find shelter and food, the plant responding to the irritation due to the presence of the larva by forming a characteristic growth, the gall, pathological but often regular and shapely, in whose hollow chamber the grub lives and eats. Ichneumon-flies and their allies pierce the skin of caterpillars and other insect-larvae, laying their eggs within the victims' bodies, which their grubs proceed to devour internally. Some very small members of these families are content to lay their eggs within the eggs of larger insects, thus obtaining rich food-supply and effective protection for their tiny larvae. In Platygaster and other genera of the family Proctotrypidae, M. Ganin (1869) showed the occurrence of hypermetamorphosis somewhat like that already described as occurring among the Oil-beetles (Meloidae). The larva of Platygaster is at first rather like a small Copepod crustacean, with prominent spiny tail-processes; after a moult this form changes into the legless grub characteristic of the Hymenoptera, among which larvae even approaching the campodeiform type are very exceptional. The species of Platygaster pass their larval stages within the larvae of gall-midges.

Wasps, bees and ants, have the ovipositor of the female modified into a sting, which is often used for the purpose of providing food for the helpless grubs. Thus the digging wasps (Sphegidae and Pompilidae) hunt for caterpillars, spiders, and other creatures which they can paralyse with their stings, and bury them alongside their eggs to furnish a food-supply for the newly-hatched young. The social wasps and many ants sting and kill flies and other insects, which they break up so as to feed their grubs within the nest. It is well known that the labour of tending the larvae in these insect societies is performed for the most part not by the mother ('Queen') but by the modified infertile females or 'workers.' Other ants and the bees feed their grubs (fig. 18), also sheltered in well-constructed nests, on honey elaborated from nectar within their own digestive canals. In all cases we see that the helplessness of the grub is associated with some kind of parental care.

From the Hymenoptera we may pass on to the Diptera or Two-winged Flies, an order of which the vast number of species and in many cases the myriads of individuals force themselves on the observer's notice. F. Brauer (1863) divided the Diptera into two sub-orders[8]; of the first of these a Crane-fly or 'Daddy-long-legs' may be taken as typical, of the second an ordinary House-fly or Bluebottle. All the larvae of the Diptera are legless, those of the Crane-fly group have well-developed hard heads, with biting mandibles, but in the House-fly section the larva is of the degraded vermiculiform type known as the maggot, not only legless, but without a definite head, the front end of the creature usually tapering to the mouth, where there are a pair of strong hooks, used for tearing up the food. A few examples of each of these types must suffice in the present brief survey. A few pages back (p. 66) reference was made to the production of galls on various plants, through the activity of larvae of the hymenopterous family Cynipidae. Many plant-galls are due, however, to the presence of grubs of tiny dipterous insects, the Cecidomyidae or Gall-midges. A cecid grub (fig. 19) has an elongate body with flexible, wrinkled cuticle, tapering somewhat at the two ends. The head, if rather narrow, is distinct, and beneath the prothorax is a characteristic sclerite known as the 'anchor process' or 'breast bone.' Along either side of the body is a series of paired spiracles, each usually situated at the tip of a little tubular outgrowth of the cuticle; the hindmost spiracles are often larger than the others. These little grubs live in family communities, their presence leading to some deformation of the plant that serves to shelter them. A shrivelled fruit or an arrested and swollen shoot, such as may be due respectively to the Pear-midge (_Diplosis pyrivora_) or the Osier-midge (_Rhabdophaga heterobia_), is a frequent result of the irritation set up by these little grubs. In a larva of the crane-fly family (Tipulidae, fig. 20) living underground and eating plant-roots, like the well-known 'leather-jacket' grubs of the large 'Daddy-long-legs' (Tipula) or burrowing into a rotting turnip or swollen fungus, like the more slender grub of a 'Winter Gnat' (Trichocera), the student notices a somewhat tough cuticle, a relatively small but distinct head, and frequently prominent finger-like processes on the tail-segment. Further examination shows a striking modification in the arrangement of the spiracles. Instead of a paired series on most of the body-segments, as in caterpillars and the vast majority of insects whether larval or adult, there are two large spiracles surrounded by the prominent tail-processes, and a pair of very small ones on the prothorax, the latter possibly closed up and useless. This restriction of the breathing-holes to a front and hind pair (amphipneustic condition) or to a hind pair only (metapneustic type) is highly characteristic of the larvae of Two-winged flies.

[8] Known as the Orthorrhapha and the Cyclorrhapha; these terms are derived from the manner in which the larval or pupal cuticle splits, as will be explained in the next chapter (p. 88).

Turning now to the maggot, characteristic of the House-fly section (fig. 21) of the Diptera, we see the greatest contrast between the larva and the imago that can be found throughout the whole class of the insects. The Bluebottle's eggs, the well-known 'fly blow' laid in summer time on exposed meat, not unnaturally arouse feelings of disgust, yet they are the prelude to one of the most marvellous of all insect life-stories. The fly--with its large globular head, bearing the extensive compound eyes, the highly modified feelers with their exquisitely feathered slender sensory tips, and the complex suctorial jaws; with its compact thorax bearing the glassy fore-wings alone used for flight, though the hind-wings modified into tiny drumstick-like 'halters' are the organs of a fine equilibrating sense--is perhaps the most specialised, structurally the 'highest' of all insects. Yet in a week or two this swift, alert, winged creature is developed from the degraded maggot, white, legless, headless, that buries itself in putrid flesh, 'feeding on corruption.'

The broad end of the maggot is the tail, while the narrow extremity marks the position of the mouth. Above this are a pair of very short feelers (fig. 21 _c_), while from the aperture project the tips of the mouth-hooks (fig. 21 e, _f_), formidable, black, claw-like structures, articulated to the strong pharyngeal sclerites and moved by powerful muscles, tearing up the fibres of the flesh. On either side of the prothorax is an anterior spiracle, a curious branching or fan-like outgrowth (fig. 21 _b_), with a variable number of tiny openings which are probably of little use for the admission of air to the tubes.

In many maggots the mouth-hooks and the front spiracles become more and more complex in form in the successive instars. The cuticle, white and smooth to the unaided eye, is seen on microscopic study to be set with rows of tiny spines which assist the maggot's movements through its food-mass. At the tail-end the large hind spiracles are conspicuous on a flattened dorsal area of the ninth abdominal segment; each shows a hard brown plate, traversed by three slits. And as we watch this curious degraded larva thrusting its narrow head-end into the depths of its ofttimes loathsome food-supply, we understand the advantage of access to the air-tube system being mainly confined to the hinder end of the body.

Maggots, differing from that of the Bluebottle only in minor details, are the larval forms of a vast multitude of allied species and display great variation in the nature of their food. Most, however, hide their soft defenceless bodies in some substance which affords shelter as well as food. The Bluebottle maggot burrows into flesh, that of the House-fly into horse-dung or vegetable refuse. The maggot of the Cabbage-fly eats its way into the roots of cruciferous plants, that of the Mangel-fly works out a broad blister between the two skins of a leaf, into which the newly-hatched larva crawls directly from the egg. A large number of species, forming an entire subfamily (the Tachininae) have larvae that feed as parasites within the bodies of other insects.

The habit of parasitism by maggots in back-boned animals has led to some remarkable modifications of the larva and to curious adventures in the course of the life-story. The Bot-fly of the Horse (_Gastrophilus equi_) and the Warble-fly of the Ox (Hypoderma bovis, fig. 22) lay eggs attached to the hairs of grazing animals, which, at least in the case of Gastrophilus, lick the newly-hatched larvae into their mouths. The 'bot,' or maggot of Gastrophilus, comes to rest in the horse's stomach; often a whole family attach themselves by their mouth-hooks to a small patch of the mucous coat of that organ. The maggot is relatively short and stout, with rows of strong spicules surrounding the segments, and with spiracles capable of withdrawal through a cup-like inpushing of the tail-region of the body, so that the parasite is preserved from drowning when the host drinks water. The young maggot of Hypoderma (fig. 22 _e_) is elongate and slender, spends its first two stages burrowing in the gullet wall and then wandering through the dorsal tissues of its host; ultimately it arrives beneath the skin of the back and assumes for its third and fourth instars a broad barrel-like form (fig. 22 _b_). The supply of free oxygen within the ox's tissues being now insufficient, the warble-maggot bores a circular hole through the skin and rests with the tail spiracles directed upwards towards the outer air. When fully grown the maggot works its way through the hole in the host's skin, and falling to the ground pupates in some sheltered spot, the life cycle occupying about a year. Similarly the Horse-bot escapes from the host's intestine with the excrement, and pupates on the ground.

A curious modification of the maggot is noticeable in the larva of the Hover-flies (Syrphus). These, unlike most of their allies, live exposed on the foliage of plants, where they feed by preying on aphids.

In agreement with this manner of life, the cuticle is roughly granulated, often greenish or reddish in hue, and the maggot, despite its want of definite head and sense organs, moves actively and purposefully about, often rearing up on its broad tail-end with an aphid victim impaled on its mouth-hooks.

In a previous chapter reference was made to the exopterygote insects, stone-flies, dragon-flies, and may-flies, whose preparatory stages live in the water. Among the endopterygote orders many Neuroptera and Coleoptera, all Trichoptera, a very few Lepidoptera and many Diptera, have aquatic larvae. One or two examples of the adaptations of dipteran larvae to life in the water may well bring the present chapter to a close. Many members of the hover-fly family (Syrphidae) have maggots with the tail-spiracles situated at the end of a prominent tubular process. Among the best-known of syrphid flies are the drone-flies (Eristalis), often seen hovering over flowers, and presenting a curious likeness to hairy bees. The larva of Eristalis is one of the most remarkable in the whole order, the 'Rat-tailed maggot' found in the stagnant water of ditches and pools. It has a cylindrical body with the hinder end drawn out into a long telescopic tube, a more slender terminal section being capable of withdrawal into, or protrusion from, a thicker basal portion. At the extremity of the slender tube is a crown of sharp processes, forming a stellate guard to the spiracles. These processes can pierce the surface-film of the water, and place the tracheal system of the maggot in touch with the pure upper air; while its mouth may be far down, feeding among the foul refuse of the ditch, it can still reach out to the medium in which the end of its life-story must be wrought out.

Reverting to the first great division of the Diptera, we find varied adaptations to aquatic life among many grubs that possess a definite head. The larva of a Gnat (Culex[9]) has projecting from the hind region of the abdomen a long tubular outgrowth, at the end of which are the spiracles, guarded by three pointed flaps forming a valve. When closed these pierce the surface-film of the water in which the larva lives; when opened a little cup-like depression is formed in the surface-film, from which the larva hangs. Or having accumulated a supply of air, it can disengage itself from the surface-film and dive through the water, its tracheal system safely closed. Another mode of breathing is found in the 'Blood-worms' and allied larvae of the Harlequin-midges (Chironomidae) whose transformations are described in detail by Miall and Hammond (1900). These larvae have two pairs of cylindrical, spine-bearing pro-legs--one on the prothorax and the other on the hindmost abdominal segment; the latter structures serve to fix the larva in the muddy tube which it inhabits at the bottom of its native pond. The penultimate abdominal segment has four long hollow outgrowths, which contain blood, and have the function of gills, while the hindmost segment has four shorter outgrowths of the same nature. Enabled thus to breathe dissolved air, the Chironomus larva needs not, like the Culex or the Eristalis, to find contact with the atmosphere beyond the surface-film.

[9] See Frontispiece, A.

Most remarkable, in many respects, of all aquatic larvae are the grubs of the Sand-midges (Simulium). These live entirely submerged and, having no special gills, carry out an exchange of gases through the general surface of the cuticle between the dissolved air in the water and the cavities of the air-tube system. The body is shaped like a flask swollen slightly at the hinder end and possesses a median pro-leg just behind the head, also another at the tail, which serves to attach the larva to a stone or to the leaf of an aquatic plant. The head has, in addition to feelers and jaws, a pair of processes with wonderful fringes which by their motion set up currents in the water, and bring food particles within reach of the mouth. A number of the larvae usually live in a community. Their power of spinning silken threads by which they can work their way back when accidentally dislodged from their resting-place, has been vividly described by Miall (1895).

Examples might be multiplied, but enough have been given to enforce the conclusion that the forms of insect-larvae are wondrously varied, and that frequently, within the limits of the same order or even family, modifications of type may be found which are suited to various modes of life adopted by different insects. A survey of the multitudes of insect larvae--grubs, caterpillars, maggots--living on land, on plants, underground, in the water; feeding on leaves, in stems, on roots, on carrion, on refuse; by hunting or by lurking after prey; as parasites or as scavengers, brings home to us most strongly the conclusion that each larva is fitted to some little niche in the vast temple of life, each is specially adapted to its part in the great drama of being.

CHAPTER VII

PUPAE AND THEIR MODIFICATIONS

The pupal stage is characteristic of the life-story of those insects whose larvae have wing-rudiments in the form of inpushed imaginal discs, and in all these insects there is, as we have seen, considerable divergence in form between larva and imago. In the pupa the wings and other characteristically adult structures are, for the first time, visible outwardly; it is the instar which marks the great crisis in transformation. The pupa rests, as a rule, in a quiescent condition, and during the early period of this stage the needful internal changes, the breaking down of many larval tissues, and their replacement by imaginal organs, go on. Both outwardly and inwardly therefore, the insect undergoes, at the pupal stage, a reconstruction necessitated by the differences in form and often in habit, between the larva and the winged adult; and the greater these differences, the more profound must be the changes that mark the pupal stage.

From the prominence of imaginal structures in the pupa, it is at once seen that the pupa of any insect must resemble the adult more nearly than it resembles the larva. But in different groups of insects we find different degrees of likeness between pupa and imago. In a beetle pupa (see fig. 16 _c_), the appendages--feelers, jaws, legs, wings--stand out from the body as do those of the perfect insect. This type is called a free pupa. The pupal cuticle has to be shed for the emergence of the imago, but the pupa is already a somewhat reduced model of the final instar, with abbreviated wings and doubled-up legs. A free pupa is characteristic of the Coleoptera,

Neuroptera, Trichoptera, Hymenoptera and many Diptera. In some cases the pupa requires to be specially adapted for a peculiar mode of life; for example, a special arrangement of breathing organs may be necessary for life under water, and there must needs be temporary pupal structures, not represented in the imago.

On the other hand, in the pupae of most Lepidoptera and of some Diptera, there is more or less coalescence between the cuticle of the appendages and the cuticle of the body generally, so that the appendages do not stand out, but being, as it were, glued down to the body, are somewhat masked (see fig. 1 e and fig. 23). Consequently the obtect pupa, as this type is called, does not resemble its imago as fully as a free pupa does. The outline of the wings for example in a butterfly's pupa can in some cases be traced only with difficulty. T.A. Chapman has shown (1893) that the completely obtect pupa characterises the more highly developed families of Lepidoptera, while in the more primitive families the pupa is incompletely obtect. If the pupa of a butterfly or moth be lifted and held in the hand, a bending or wriggling motion of the abdomen can be observed. In the incompletely obtect pupa, this motion is evident in a greater number of segments than in the completely obtect, the number concerned varying from five to two in different families. In the nymphalid butterflies, the pupa is often called a 'chrysalis' on account of the golden hue displayed by the cuticle, and the term 'chrysalis' is sometimes bestowed indiscriminately on any kind of pupa. It has been shown by Poulton (1892) and others, that the colour of a butterfly pupa is to some extent affected by the surroundings of the caterpillar just before its last moult.

Reference has been made (p. 58) to the power of spinning silk possessed by many larvae; often the principal use of this silk is to form some protection for the pupa, the larva before its last moult constructing a cocoon within which the pupa may rest safely. Many larvae bury themselves in the earth, and the pupa lies in an earthen chamber, the lining particles of soil fastened together by fine silken threads. Larvae that feed in wood, like the caterpillar of the Goat-moth (Cossus) make a cocoon of splinters spun together, while hairy caterpillars, such as those of the Tiger-moths, work some of their hairs in with the silk to make a firm cocoon (fig. 17 _b_). On the other hand, those caterpillars known as 'silkworms' make a dense cocoon of pure silk, consisting of two layers, the outer of coarse and the inner of fine threads. Silken cocoons very similar in appearance are spun by the larvae of small Ichneumon-flies. Many pupae lie in a loose cocoon formed of a few interlacing threads, as for example the conspicuous black and yellow banded pupa of the Magpie-moth (_Abraxas grossulariata_) and the pupae of various leaf-beetles. Others again spin together the edges of leaves with connecting silken threads. The grubs of bees and wasps which are reared in the comb-chambers of their nests seal up the opening of the chamber with a lid, partly silk (fig. 18 _co_) and partly excretion, when ready to pass into the pupal state. An additional external 'capping' may be also supplied by the workers.

The pupae of butterflies are especially interesting, as illustrating the extreme reduction of the silken cocoon. The pupa of a 'swallowtail' (Papilionid) or a 'white' (Pierid) butterfly (fig. 23) may be found attached to a twig of its food-plant or to a wall, in an upright position, its tail fastened to a pad of silk and a slender silken girdle encircling its thorax. The pupa of a 'Tortoiseshell' or 'Admiral' (Nymphalid) butterfly hangs head downwards from a twig, supported only by the tail-pad of silk, which, useless as a shelter, serves only for attachment. The pupa is fastened to this pad by a spiny hook or process, the cremaster (fig. 23 _cr_), on the last abdominal segment. The cremaster is a characteristic structure in the pupa of a moth or butterfly. C.V. Riley (1880) and W. Hatchett-Jackson (1890) have shown that it corresponds with a spiny area, the suranal plate, which lies above the opening of the caterpillar's intestine. The means by which the suspended pupa of a nymphalid butterfly attaches its cremaster to the silken pad which the larva has spun in preparation for pupation, is worthy of brief attention. The caterpillar, hanging head downwards, is attached to the silken pad by its hindmost pair of pro-legs or claspers and by the suranal plate, and the cuticle is slowly worked off from before backwards, so as to expose the pupa. Were the process of moulting to be simply completed while the insect hangs by the claspers, the pupa would of course fall to the ground. But there is enough adhesion between the pupal and larval cuticles at the hinder end of the body, especially by means of the everted lining of the hind-gut, for the pupa to be supported while it jerks its cremaster out of the larval cuticle and works it into the meshes of the silken pad. The moult is thus completed and the pupa hangs securely all the time. In the numerous cases where the pupa is enclosed in a cocoon, the cremaster serves to fix the pupa to the surrounding silk. Chapman (1893) has drawn attention to the fact that among the more highly organised moths the pupa remains in the cocoon, the emergence being entirely left to the imago, while the pupae of the more primitive moths work their way partly out of the cocoon before the final moult begins. In the latter case, the cremaster is anchored by a strand of silk which allows a certain degree of emergence, and the pupa has rows of spines on its abdominal segments, of which a greater number retain the power of mutual motion than in those pupae which do not come out of their cocoons.

While the pupa on the whole resembles the imago that is to emerge from it, there are not a few cases in which a special structure necessary for some contingency in pupal life is retained or adopted in this stage. A butterfly pupa, like the imago, has no mandibles, but in the case of the Caddis-flies (Trichoptera) and two families of small moths, the most primitive of all Lepidoptera, the pupa, like the larva, has well-developed mandibles. These enable the caddis pupa to bite its way out of the shortened larval case in which it has pupated, and then to swim upwards through the water ready for the caddis-fly's emergence into the air. Pupae that are submerged require special breathing-organs. In the previous chapter (p. 77) mention was made of the gnat's aquatic larva with its tail-spiracles adapted for procuring atmospheric air through the surface-film. The pupa of the gnat[10] also has 'respiratory trumpets' serving the same purpose, but these are a pair of processes on the prothorax, so that the pupa, which is fairly active, hangs from the surface-film with its abdomen pointing downwards through the water. This change of position is correlated with the necessity for the imago to emerge into the air; were the pupa to hang head downwards as the larva does, the gnat would perforce have to dive into the water. With the beautifully adapted transfer of the functional spiracles, their position is appropriately arranged for the gnat's emergence at the surface, and the empty pupal cuticle floats serving the insect as a raft. On this it rests securely and the crumpled wings have opportunity to expand and harden before the insect takes to flight.

[10] See Frontispiece, B.

The aquatic pupae of other Diptera, many species of the midges Chironomus and Simulium for example, breathe dissolved air by means of tufts of thread-like gills, which arise on either side of the prothorax. The pupae of Simulium rest in their curious little cup-like dwellings, attached to submerged stones or plants. The Chironomus pupa is usually found in an elongate gelatinous case adhering to a stone. From this case the pupa rises to the surface of the water, that the midge may emerge into the air. Miall and Hammond (1900) describe the arrangement by which, when the pupal stage ends, and these gills are no longer required, their connection with the air-tube system is severed 'without undue violence.' The walls of the fine air-tubes that pass into the gills are specially strengthened, but just below the pupal cuticle these walls are exceedingly thin and delicate. Thus when the pupal cuticle is cast, they are readily broken there, and the cuticle of the midge forming beneath has a spiracular opening into the main air-trunk, ready for use during the insect's aerial life.

Among those Diptera whose larva is the headless maggot a most remarkable arrangement for protecting the pupa is to be found. The last larval cuticle, instead of being as usual worked off and cast, after separation from the underlying structures, becomes hard and firm, forming a protective case (_puparium_) within which by the processes of histolysis and histogenesis already described the organs of the pupa and imago are built up. This puparium (fig. 22 _d_) is usually dark in colour, often brown and barrel-shaped, and a subcircular lid splits off from it at the head-end to allow the emergence of the fly[11]. While the maggot breathes by its tail-spiracles, the functional spiracles of the puparium (connected with the tracheal system of the enclosed pupa) are far forward, and these may be situated at the tips of long sometimes branching processes, which recall the thoracic gills of the aquatic pupae mentioned a few pages above. Adaptations, various and beautiful, to special modes of life, are thus seen to characterise pupae as well as larvae.

[11] The presence of this sub-circular lid characterises Brauer's suborder Cyclorrhapha. Those Diptera in which the pupal cuticle splits in the normal, longitudinal manner are included in the Orthorrhapha (see p. 67).

CHAPTER VIII

THE LIFE-STORY AND THE SEASONS

A number of interesting questions are associated with the seasonal cycle of an insect's life-history. In a previous chapter (IV. pp. 30, 34) reference has been made to the contrast between the long aquatic life of the larval dragon-fly or may-fly, extending over several years, and the short aerial existence of the winged adult restricted in the case of the may-flies to a few hours. Here we see that the feeding activities of the insect are carried on during the larval stage only; the may-fly in its winged condition takes no food, pairing and egg-laying form the whole of its appointed task. A similar though less extreme shortening of the imaginal life may be noticed in many endopterygote insects. For example, the bot- and warble-flies have the jaws so far reduced that they are unable to feed, and the parasitic life of the maggot (see p. 74) extending over eight or nine months in the body of the horse or ox, prepares for a winged existence of probably but a few days. Again in many moths the jaws are reduced or vestigial so that no food can be taken in the winged state, as for example in the 'Eggars' (Lasiocampidae) and the 'Tussocks' (Lymantriidae). It is noteworthy that in these short-lived insects the male is often provided with elaborate sense-organs which, we may believe, assist him to find a mate with as little delay as possible; the male may-fly has especially complex eyes, while the feelers of the male silk-moth or eggar are comb-like or feathery, the branches bearing thousands of sensory hairs. A box with a captive living female of one of these moths, if taken into a wood haunted by the species becomes rapidly surrounded by a swarm of would-be suitors, attracted by the odour emitted from the prisoner's scent-glands.

Very exceptionally the imaginal stage may be omitted from the life-story altogether. Nearly fifty years ago N. Wagner (1865) made the remarkable discovery that in the larvae of certain gall-midges (Cecidomyidae) the ovaries might become precociously mature and unfertilised eggs might be developed into small larvae observable within the body of the mother-larva; ultimately these abnormally reared young break their way out. In this case therefore there may be a series of larval generations, neither pupa nor imago being formed. Extended observations on the precocious reproductive processes of these midges have lately been published by W. Kahle (1908). A less extreme instance of an abbreviated life-story was made known by O. Grimm (1870) who saw pupae of Harlequin-midges (Chironomus) lay unfertilised eggs, which developed into larvae. Here the imaginal stage only is omitted from the life-history. Not always however is it the imaginal stage of the life-history which is shortened. Reference (p. 18) has already been made to the case of the virgin female aphids, whose eggs develop within the mother's body, so that active, formed young are brought forth. Among the Diptera it is not unusual to find similar cases, the female fly giving birth to young maggots instead of laying eggs. Such is the habit of the great flesh-fly (Sarcophaga), of some allied genera (Tachina, etc.) whose larvae live as parasites on other insects, and occasionally of the Sheep Bot-fly (Oestrus). In such cases we recognise the beginning of a shortened larval period, and Brace's investigations in 1895, summarised by E.E. Austen (1911), have shown that females of the dreaded African Tsetse flies (Glossinia) bring forth nearly mature larvae, which pupate soon after birth. In another group of Diptera, the blood-sucking parasites of the Hippoboscidae and allied families, the whole larval development is passed through within the mother's body, and a full-grown larva is born the cuticle of which hardens and darkens immediately to form a puparium; hence these flies are often called, though incorrectly, Pupipara. Still more astonishing is the mode of reproduction in the allied family of the Termitoxeniidae, curious, degraded, wingless 'guests' of the termites, or 'white ants,' lately made known through the researches of E. Wasmann (1901). Here the individual is hermaphrodite--a most exceptional condition among insects--and lays a large egg, whence is usually hatched a fully-developed adult! Here then we find that all the early stages, usual in the higher insects, are omitted from the life-story.

Interesting comparison may be made between the total duration of various insect life-stories. To some extent at least, the length of an insect's life is correlated with its size, its food, the season of the year when it breeds. Small insects have, as a rule, shorter lives than large ones; those whose larvae devour highly nutritive food generally develop more quickly than those which have to live on dry, poor, substances; life-cycles follow one another most rapidly in summer weather when temperature is high and food plentiful.

In early chapters we have already noticed the long aquatic life of the larva and nymph of a dragon-fly, relatively a large insect, and the rapid multiplication of the repeated summer broods of virgin aphids (p. 18). Within the one order of the Coleoptera it is instructive to compare the small jumping leaf-beetles, the 'turnip-flies' of the farmer, whose larvae mine in the green tissues, and complete their transformations so rapidly that several successive broods appear in the spring and early summer, with the larger click-beetles whose larvae, the equally notorious 'wireworms,' feed on roots for three or four years before they become fully grown. Among the Diptera, the 'leather-jacket' grub of the crane-fly, feeding like the wireworm on roots, has a larval life extending through the greater part of a year, while the maggot of the bluebottle, feeding on a rich meat diet, becomes mature in a few days. As examples of excessively long life-cycles the 'thirteen-year' and 'seventeen-year' cicads of North America, described by C.L. Marlatt (1895), are noteworthy. Certain specially populous 'broods' of these insects are known and localised, so that the appearance of the imagos in future years can be accurately predicted. Here again we have to do with bulky insects whose subterranean larvae and nymphs feed on comparatively innutritious roots.

In our own climate, it is of interest to notice the variation among insects as to the stage which carries the race over the winter. The click-beetles, mentioned just above, emerge from their buried pupae in summer, hibernate under stones or clods, and lay eggs among the herbage next spring. At the same time of course, owing to the extended term of the larval life, many more individuals of the species are wintering underground as 'wireworms' of various ages, and these, except in very severe frosts, can continue their occupation of feeding on roots. But in the case of the 'turnip-flies' the food-supply is cut off in winter, and all those beetles of the latest summer brood that survive hibernate in some sheltered spot, waiting for the return of spring, that they may lay their eggs, and start the life-cycle once again. Among the Diptera, most species pass the winter as pupae, the sheltering puparium being a good protection against most adverse conditions, or as flies. But where there is a prolonged parasitic larval life, as with the bot- and warble-flies, the maggot, warm and well-fed within the body of its mammalian host, affords an appropriate wintering stage.

Among the Hymenoptera an especially interesting seasonal life-cycle is afforded by the alternation of summer and winter generations in many Gall-flies (Cynipidae) as H. Adler (1881, 1896) demonstrated for most of our common species. The well-known 'oak-apples' are tenanted in summer by grubs, which after pupation develop into winged males and wingless females. The latter, after pairing, burrow underground and lay their eggs in the roots, the larvae causing the presence there of globular swellings or root-galls within which they live, pass through their transformations and develop into wingless virgin females. These shelter until February or March in their underground chambers, then climb up the tree and lay on the shoots eggs, from which will be hatched the grubs destined to grow within the oak-apples into the summer sexual brood of flies.

The Lepidoptera afford examples of hibernation in all stages of the life-history. In this order a few large moths with wood-boring caterpillars, the 'Goat' (Cossus) for example, undergo a development extending over several years, while at the other extreme a few small species may have three or more complete cycles within the twelve months. But in the vast majority of Lepidoptera we find either one or two generations, definitely seasonal, within the year; the insect is either 'single-brooded' or 'double-brooded.'

Almost every winter one or more letters may be read in some newspaper recording the writer's surprise at seeing on a sunny day during the cold season, one of our common gaily-coloured butterflies of the Vanessa group, a 'Tortoiseshell' or 'Red Admiral,' flitting about. Surprise might be greater did the observers realise that the imaginal is the normal hibernating stage for these species. Emerging from the pupa in late summer or autumn, they shelter during winter in hollow trees, under thatched eaves, in outbuildings or in similar situations, coming out in spring to lay their eggs on the leaves of their caterpillars' food-plants. The larvae feed and grow through the early summer months, in the case of the Small Tortoiseshell (_Vanessa urticae_) pupating before midsummer and developing into a July brood of butterflies whose offspring after a late summer life-cycle, hibernate; while for the larger species of the group there is, in our islands, only one complete life-cycle in the year, though the same insects in warmer countries may be double-brooded. C.G. Barrett records (1893, vol. I. pp. 153-4) how in the August of 1879 hundreds and thousands of 'Painted Ladies' (_Pyrameis cardui_) migrated into the south of England from the European continent where in many places great swarms had been observed early in the summer. 'These August butterflies, the progeny of the June swarms, coming from a warmer climate, had no intention of hibernating, but paired and laid eggs. Some of the larvae were collected and reared indoors [butterflies] emerging in November and December, but out of doors all must have been destroyed by damp or frost, in either the larva or pupa state, for no freshly emerged specimens were noticed in the spring, and no trace of the great migration remained.'

In September and October the pedestrian, even in a suburban square, may see moths with pretty brown, white-spotted wings flying around trees. These are males of the common 'Vapourer' (_Orgyia antiqua_), in search of the females which, wingless and helpless, rest on the cocoons surrounding the pupae whence they have just emerged, the cocoons being attached to the branches of the trees where the caterpillars have fed. After pairing, the female lays her eggs among the silk of the cocoon, partly covering them with hairs shed from her body, and then dies. The eggs thus protected remain through the winter, the larvae not being hatched till springtide, when the young leaves begin to sprout forth. The caterpillars, adorned and probably protected by their 'tussocks' of black or coloured bristles, feed vigorously. Their activity and habit of occasional migration from one tree to another, compensates, to some extent, as Miall (1908) has pointed out, for the females' enforced passivity; only in the larval state can moths with such wingless females extend their range. The caterpillars spin their cocoons towards the end of summer, and then pupate, the moths emerging in the autumn and the eggs, as we have seen, furnishing the winter stage.

After midsummer, the conspicuous cream, black and yellow-spotted 'Magpie' moth (_Abraxas grossulariata_) is common in gardens. The female lays her eggs on a variety of shrubby plants; gooseberry and currant bushes are often chosen. From the eggs caterpillars are hatched in autumn, but these, instead of beginning to feed, seek almost at once for rolled-up leaves, cracks in walls, crannies of bark, or similar places, which may afford winter shelters. Here they remain until the spring, when they come out to feed on the young foliage and grow rapidly into the conspicuous cream, yellow and black 'looper' caterpillars mentioned in a previous chapter (p. 60). These, when fully-grown, spin among the twigs of the food-plant a light cocoon, in which the black and yellow-banded wasp-like pupa spends its short summer term before the emergence of the moth.

An equally familiar garden insect, the common 'Tiger' moth (_Arctia caia_) with its 'woolly bear' caterpillar, affords a life-cycle slightly differing from that of the 'Magpie.' The gaudy winged insects are seen in July and August, and lay their eggs on a great variety of plants. The larvae hatched from these eggs begin to feed at once, and having moulted once or twice and attained about half their full size, they rest through the winter, the dense hairy covering wherewith they are provided forming an effective protection against the cold. At the approach of spring they begin to feed again, and the fully-grown 'woolly bear' is a common object on garden paths in May and June. Before midsummer it has usually spun its yellow cocoon under some shelter on the ground and changed into a pupa.

Another modification with respect to seasonal change is shown by the Turnip moth (_Agrotis segetum_) and other allied Noctuidae (Owl-moths). These are insects with brown-coloured wings, flying after dark in June. The dull greyish larvae feed on many kinds of low-growing plants, usually hiding in the earth by day and wandering along the surface of the ground by night, biting off the farmer's ripening corn, or burrowing into his turnips or potatoes. On account of the burrowing habits of this insect it can feed throughout the winter, except when a hard frost puts a temporary stop to its activity. By April it has become fully grown and pupates in an earthen chamber a few inches below the surface. The Turnip moth in our countries is partially double-brooded, a minority of the autumn caterpillars growing more rapidly than their comrades so that they pupate, and a second brood of moths appear in September. These pair and lay eggs, the resulting caterpillars going as Barrett suggests (1896, vol. III. p. 291) 'to reinforce the great army of wintering larvae.'

Such underground caterpillars, to a great extent protected from cold, can continue to feed through the winter. With other species we find that the larva becomes fully grown in autumn, yet lives through the winter without further change. This is the case with the Codling moth (_Carpocapsa pomonella_), a well-known orchard pest, which in our countries is usually single-brooded. The moth is flying in May and lays her eggs on the shoots or leaves of apple-trees, more rarely on the fruitlets, into which however the caterpillar always bores by the upper (calyx) end. Here it feeds, growing with the growth of the fruit, feeding on the tissue around the cores, ultimately eating its way out through a lateral hole, and crawling upwards if its apple-habitation has fallen, downwards if it still remains on the bough, to shelter under a loose piece of bark where it spins its cocoon about midsummer and hibernates still in the larval condition. Not until spring is the pupal form assumed, and then it quickly passes into the imaginal state. In the south of England, as F.V. Theobald (1909) has lately shown, and also in southwestern Ireland, this species may be double-brooded, the usual condition on the European continent and in the United States of America. There the midsummer larvae pupate at once and the moths of an August brood lay eggs on the hanging or stored fruit; in this case, again, however, the full-grown larva, quickly fed-up within the developed apples, is the wintering stage.

Several of the insects mentioned in this survey, like the last-named codling moth, are occasionally double-brooded. As an example of the many Lepidoptera, which in our islands have normally two complete life-cycles in the year, we may take the very familiar White butterflies (Pieris) of which three species are common everywhere. The appearance of the first brood of these butterflies on the wing in late April or May is hailed as a sign of advanced spring-time. They pair and lay their eggs on cabbages and other plants, and the green hairy caterpillars feed in June and July, after which the spotted pupae may be found on fences and walls, attached by the silken tail-pad and supported by the waist-girdle. In August and September butterflies of the second brood have emerged from these and are on the wing; their offspring are the autumn caterpillars which feed in some seasons as late as November, doing often serious damage to the late cruciferous crops before they pupate. The pupae may be seen during the winter months, waiting for the spring sunshine to call out the butterflies whose structures are being formed beneath the hard cuticle.

Reviewing the small selection of life-stories of various Lepidoptera just sketched, we notice an interesting and suggestive variety in the wintering stage. The vanessid butterflies hibernate as imagos; the 'vapourer' winters in the egg, the magpie as a young ungrown larva, the 'tiger' as a half-size larva; the Agrotis caterpillar feeds through the winter, growing all the time; the codling caterpillar completes its growth in the autumn, and winters as a full-size resting larva; lastly, the 'whites' hibernate in the pupal state. And in every case it is noteworthy that the form or habit of the wintering stage is well adapted for enduring cold.

Our native 'whites' afford illustration of another interesting feature often to be noticed in the life-story of double-brooded Lepidoptera. The butterflies of the spring brood differ slightly but constantly from their summer offspring, affording examples of what is called seasonal dimorphism. All three species have whitish wings marked with black spots, larger and more numerous in the female than in the male. In the spring butterflies these spots tend towards reduction or replacement by grey, while in the summer insects they are more strongly defined, and the ground colour of the wings varies towards yellowish. In the 'Green-veined' white (_Pieris napi_) the characteristic greenish-grey lines of scaling beneath the wings along the nervures, are much broader and more strongly marked in the spring than in the summer generation, whose members are distinguished by systematic entomologists under the varietal name napaeae. The two forms of this insect were discussed by A. Weismann in his classical work on the Seasonal Dimorphism of butterflies (1876). He tried the effect of artificially induced cold conditions on the summer pupae of Pieris napi, and by keeping a batch for three months at the temperature of freezing water, he succeeded in completely changing every individual of the summer generation into the winter form. The reverse of this experiment also was attempted by Weismann. He took a female of bryoniae, an alpine and arctic variety of Pieris napi, showing in an intensive degree the characters of the spring brood. This female laid eggs the caterpillars from which fed and pupated. The pupae although kept through the summer in a hothouse all produced typical bryoniae, and none of these with one exception appeared until the next year, for in the alpine and arctic regions this species is only single-brooded. Weismann experimented also with a small vanessid butterfly, Araschnia levana, common on the European continent, though unknown in our islands, which is double (or at times treble) brooded, its spring form (_levana_) alternating with a larger and more brightly coloured summer form (_prorsa_). Here again by refrigerating the summer pupae, butterflies were reared most of which approached the winter pattern, but it was impossible by heating the winter pupae to change levana into prorsa. Experiments with North American dimorphic species have given similar results. Weismann argued from these experiments that the winter form of these seasonally dimorphic species is in all cases the older, and that the butterflies developing within the summer pupae can be made to revert to the ancestral condition by repeating the low-temperature stimulus which always prevailed during the geologically recent Ice Age. On the other hand, a high temperature stimulus applied to one generation of the winter pupae cannot induce the change into the summer pattern, which has been evolved still more recently by slow stages, as the continental climate has become more genial. In tropical countries where instead of an alternation of winter and summer, alternate dry and rainy seasons prevail, somewhat similar seasonal dimorphism has been observed among many butterflies. Not a few forms of Precis, an African and Indian genus allied to our Vanessa, that had long been considered distinct species are now known, thanks to the researches of G.A.K. Marshall (1898), to be alternating seasonal forms of the same insect. The offspring when adult does not closely resemble the parent; its appearance is modified by the climatic environment of the pupa. The experiments of Weismann just sketched in outline show at least that the same principle holds for our northern butterflies.

We are thus led to see from the life-story of such insects, that the course of the story is not rigidly fixed; the creature in its various stages is plastic, open to influence from its surroundings, capable of marked change in the course of generations. And so the seasonal changes in the history of the individual from egg to imago point us to changes in the age-long history of the race.

CHAPTER IX

PAST AND PRESENT; THE MEANING OF THE STORY

In the previous chapter we recognised how the seasonal changes in various species of butterflies as observable in two or three generations, indicate changes in the history of the race as it might be traced through innumerable generations. The endless variety in the form and habits of insect-larvae and their adaptations to various modes of life, which have been briefly sketched in this little book, suggest vaster changes in the class of insects, as a whole, through the long periods of geological time. Every student of life, influenced by the teaching of Charles Darwin (1859) and his successors, now regards all groups of animals from the evolutionary standpoint, and believes that comparisons of facts of structure and life-history of orders and classes evidently akin to each other, furnish at least some indications of the course of development in the greater systematic divisions, even as the facts of seasonal dimorphism, mentioned in the last chapter, give hints as to the course of development in those restricted groups that we call species or varieties. A brief discussion of the main outlines of the life-story of insects in the wide, evolutionary sense may thus fitly conclude this book.

In the first place we turn to the 'records' of those rocks, in whose stratified layers[12] are entombed remains, often fragmentary and obscure, of the insects of past ages of the earth's history. Compared with the thousands of extinct types of hard-shelled marine animals, such as the Mollusca, fossil insects are few, as could only be expected, seeing that insects are terrestrial and aerial creatures with slight chance of preservation in sediments formed under water. Yet a number of insect remains are now known to naturalists, who are, in this connection, more particularly indebted to the researches of S.H. Scudder (1885), C. Brongniart (1894), and A. Handlirsch (1906).

We are now considering insects from the standpoint of their life-histories, and the individual life-story of an insect of which we possess but a few fragments of wings or body, entombed in a rock formed possibly before the period of the Coal Measures, can only be a matter of inference. Still it may safely be inferred that when the structure of these remains clearly indicates affinity to some existing order or family, the life-history of the extinct creature must have resembled, on the whole, that of its nearest living allies. And all the fossil insects known can be either referred to existing orders, or shown to indicate definite relationship to some existing group.

Passing over some doubtful remains of Silurian age, we find in rocks usually regarded as Devonian[13] the most ancient fossils that can be certainly referred to the insects, while from beds of the succeeding Carboniferous period, a number of insect remains have been disinterred. These Palaeozoic insects were frequently of large size, and they show distinct affinities with our recent may-flies, dragon-flies, stone-flies, and cockroaches. In the Permian period, the latest of the divisions of the Palaeozoic, lived Eugereon, an insect with hemipteroid jaws and orthopteroid wings. All these insects must have been exopterygote in their life-history, if we may trust the indications of affinity furnished by their structure. In the Mesozoic period, however, insects with complete transformations must have been fairly abundant. Rocks of Triassic age have yielded beetles and lacewing-flies, while from among Jurassic fossils specimens have been described as representing most of our existing orders, including Lepidoptera, Hymenoptera and Diptera. In Cainozoic rocks fossil insects of nearly six thousand species have been found, which are easily referable to existing families and often to existing genera. We may conclude then, imperfect though our knowledge of extinct insects is, that some of the most complex of insect life-stories were being worked out before the dawn of the Cainozoic era. Some instructive hints as to differences in the rate of change among different insect groups may be drawn from the study of parasites. For example, V.L. Kellogg (1913) points out that an identical species of the Mallophaga (Bird-lice) infests an Australian Cassowary and two of the South American Rheas; while two species of the same genus (Lipeurus) are common to the African Ostrich and a third kind of South American Rhea. These parasites must have been inherited unchanged by the various members of these three families of flightless birds from their common ancestors, that is from early Cainozoic times at latest. On the other hand, the various kinds of such highly specialised parasites as the warble-flies of the oxen and deer, must have become differentiated during those later stages of the Cainozoic period which witnessed the evolution of their respective mammalian hosts.

[13] The 'Little River' beds of St John, New Brunswick, Canada, by some modern geologists however considered as Carboniferous.

The foregoing brief outline of our knowledge of the geological succession of insects shows that the exopterygote preceded, in time, the endopterygote type of life-history. We have already seen that those insects undergoing little change in the life-cycle, and with visible, external wing-rudiments, are on the whole less specialised in structure than those which pass through a complete transformation. These two considerations, taken together, suggest strongly that in the evolution of the insect class, the simpler life-history preceded the more complex. Such a conclusion seems reasonable and what might have been expected, but we are confronted with the difficulty that if the most highly organised insects pass through the most profound transformations, then insects present a remarkable and puzzling exception to the general rules of development among animals, as has already been pointed out in the first chapter of this volume (p. 7). A few students of insect transformation have indeed supposed that the crawling caterpillar or maggot must be regarded as a larval stage which recalls the worm-like nature of the supposed far-off ancestors of insects generally. Even in Poulton's classical memoir (1891, p. 190), this view finds some support, and it may be hard to give up the seductive idea that the worm-like insect-larva has some phylogenetic meaning. But the weight of evidence, when we take a comprehensive survey of the life-story of insects, must be pronounced to be strongly in favour of the view put forward by Brauer (1869), and since supported by the great majority of naturalists who have discussed the subject, that the caterpillar or the maggot is itself a specialised product of the evolutionary process, adapted to its own particular mode of larval life.

The explanation of insect transformation is, in brief, to be found in an increasing amount of divergence between larva and imago. The most profound metamorphosis is but a special type of growth, accompanied by successive castings and renewings of the chitinous cuticle, which envelopes all arthropods. In the simplest type of insect life-story, there is no marked difference in form between the newly-hatched young and the adult, and in such cases we find that the young insect lives in the same way as the adult, has the same surroundings, eats the same food. This is the rule (see Chapters II and III) with the Apterygota, the Orthoptera, and most of the Hemiptera. In the last-named order, however, we find in certain families marked divergence between larva and imago, for example in the cicads, whose larvae live underground, while in the coccids, whose males are highly specialised and females degraded, there succeeds to the larva--very like the young stage in allied families--a resting instar, which in the case of the male, suggests comparison with the pupa of a moth or beetle.

Turning to the stone-flies, dragon-flies and may-flies, whose life-stories have been sketched in Chapter IV, we find that the early stages are passed in water, whence before the final moult, the insects emerge to the upper air. Except for the possession of tufted gills, adapting them to an aquatic life, the stone-fly nymphs differ but slightly from the adults; the grubs of the dragon-flies and may-flies, however, are markedly different from their parents. In connection with these comparisons, it is to be noted that the dragon-flies and may-flies are more highly specialised insects than stone-flies, divergent specialisation of the adult and larva is therefore well illustrated in these groups, which nevertheless have, like the Hemiptera and Orthoptera, visible external wing-rudiments.

From the vast array of insects that show internal wing-growth and a true pupal stage, a few larval types were chosen for description in Chapter VI, and a review of these suggests again the thought of increasing divergence between larva and imago. Reference has been made previously to the many instances in which the former has become pre-eminently the feeding, and the latter the breeding stage in the life-cycle. It seems impossible to avoid the conclusion that the active, armoured campodeiform grub differing less from its parent than an eruciform larva differs from its parent, is as a larval type more primitive than the caterpillar or maggot. A. Lameere has indeed, while admitting the adaptive character of insect larvae generally, argued (1899) with much ingenuity that the eruciform or vermiform type must have been primitive among the Endopterygota, believing that the original environment of the larvae of the ancestral stock of all these insects must have been the interior of plant tissues. He is thus forced to the necessity of suggesting that the campodeiform larvae of ground-beetles or lacewings must be regarded as due to secondarily acquired adaptations; 'they resemble Thysanura and the larvae of Heterometabola only as whales resemble fishes.' There are two considerations which render these theories untenable. The Neuroptera and Coleoptera among which campodeiform larvae are common, are less specialised than Lepidoptera, Hymenoptera, and Diptera, in which they are unknown. And among the Coleoptera which as we have seen (pp. 50 _f._) display a most interesting variety of larval structure, the legless, eruciform larva characterises families in which the imago shows the greatest specialisation, while in the same life-story, as in the case of the oil-beetles (pp. 56-7), the newly-hatched grub may be campodeiform, changing to the eruciform type as soon as it finds itself within reach of its host's rich store of food.

A certain amount of difficulty may be felt with regard to the theory of divergent evolution between imago and larva, in the case of those insects with complete transformation whose grubs and adults live in much the same conditions. By turning over stones the naturalist may find ground-beetles in company with the larvae of their own species. On the leaves of a willow tree he may observe leaf-beetles (Phyllodecta and Galerucella) together with their grubs, all greedily eating the foliage; or lady-bird beetles (Coccinella) and their larvae hunting and devouring the 'greenfly.' All of these insects are, however, Coleoptera, and the adult insects of this order are much more disposed to walk and crawl and less disposed to fly than other endopterygote insects. Their heavily armoured bodies and their firm shield-like forewings render them less aerial than other insects; in many genera the power of flight has been altogether lost. It is not surprising, therefore, that many beetles, even when adult, should live as their larvae do; since the acquirement of complete metamorphosis they have become modified towards the larval condition, and an extreme case of such modification is afforded by the wingless grub-like female Glow-worm (Lampyris).

With most insects, however, the larva must be regarded as the more specially modified, even if degraded, stage. Miall (1895) has pointed out that the insect grub is not a precociously hatched embryo, like the larvae of multitudes of marine animals, but that it exhibits in a modified form the essential characters of the adult. Comparison for example can be readily made between the parts of the caterpillar and the butterfly, whose story was sketched in the first chapter of this book, widely different though caterpillar and butterfly may appear at a superficial glance. And the survey of variety in form, food, and habit of insect larvae given in Chapter VI enforces surely the conclusion that the larva is eminently plastic, adaptable, capable of changing so as to suit the most diverse surroundings. In a most suggestive recent discussion on the transformation of insects P. Deegener (1909) has claimed that the larva must be regarded as the more modified stage, because while all the adult's structures are represented in the larva, even if only as imaginal buds, there are commonly present in the larva special adaptive organs not found in the imago, for example the pro-legs of caterpillars or the skin-gills of midge-grubs. The correspondence of parts in butterfly and caterpillar just referred to, may still be traced, though less easily, in bluebottle and maggot. The latter is an extreme example of degenerative evolution, and its contrast with the elaborately organised two-winged fly marks the greatest divergence observable between the larva and imago. With this divergence the resting pupal stage, during which more or less dissolution and reconstruction of organs goes on, becomes a necessity, and it has already been pointed out how the amount of this reconstruction is greatest where the divergence between the larval and perfect stages is most marked. Whatever differences of opinion may prevail on points of detail, the general explanation of insect metamorphosis as the result of divergent evolution in the two active stages of the life-story must assuredly be accepted. No other explanation accords with the increasing degree of divergence to be observed as we pass from the lower to the higher insect orders.

The successive incidents of the life-story of most insects are largely connected with the acquisition of wings. Wings, and the power of flight wherewith they endow their possessors, are evidently beneficial to the race in giving power of extending the range during the breeding period and thus ensuring a wide distribution of the eggs. In no case are wings fully developed until the closing stage of the insect's life, they are always acquired after hatching or birth. We have already noticed (p. 40) how Sharp (1899) has laid stress on the essential difference between the exopterygote and endopterygote insects, the wing-rudiments of the former growing outwards throughout life while those of the latter remain hidden until the pupal instar. Sharp considers that there is some difficulty in bridging, in thought, the gap between these two methods of wing-growth, and has put forward an ingenious suggestion to meet it (1902). Reference has already been made to insects of various orders in which one sex is wingless, the Vapourer Moth (p. 96) for example, or all the individuals of both sexes are wingless, as the aberrant cockroaches mentioned in Chapter II (p. 15), or certain generations of virgin females are wingless, for example aphids (pp. 18-19) and gall-flies (pp. 94-5). Insects may thus become secondarily wingless, that is to say be manifestly the offspring of winged parents, and such wingless forms may on the other hand give rise to offspring or descendants with well-developed wings. Frequently, as in the case of the aphids, many wingless generations intervene between two winged generations. A striking illustration of this fact is afforded by an aquatic bug, Velia currens, commonly to be seen skating over the surface of running water. The adults of Velia are nearly always wingless, but now and then the naturalist meets with a specimen provided with functional wings, the possession of which enables the insect to make its way to a fresh stream. Moreover there are whole orders of parasitic insects, such as the lice and fleas, which, showing clear affinity to orders of winged insects, are believed to be secondarily wingless. These orders are designated by Sharp 'Anapterygota.' And from the analogy of the periodic loss and recovery of wings in various generations of the same species, he has concluded that the gap between the exopterygote and the endopterygote method of development may have been bridged by an anapterygote condition; that the ancestors of those insects with complete transformations were the wingless descendants of primitive insects which grew their wings from visible external rudiments, and that in later times re-acquiring wings, they developed these organs in a new way, from inwardly directed rudiments or imaginal buds.

This theory of Sharp's is original, daring, and ingenious, but the loss and re-acquisition of wings which it presupposes is difficult to imagine in large groups during a prolonged evolutionary history, while the sudden appearance of a totally new mode of wing-growth in the offspring of wingless insects would be an extreme example of discontinuity in development.

On the whole the most probable suggestion which can be made as to the origin of 'complete' transformation in insects is that the instar in which wings were first visible externally became later and later in the course of the evolution of the more highly organised groups. In this way a gradual transition from the exopterygote to the endopterygote type of life-story is at least conceivable. It will be remembered that a may-fly (p. 33) undergoes a moult after acquiring functional wings, emerging into the air as a 'sub-imago.' In not a few endopterygote insects, the pupa shows more or less activity, swimming through water intermittently (gnats) or just before the imago has to emerge (caddis-flies); working its way out of the ground (crane-flies) or coming half-way out of its cocoon (many moths). The pupa of the higher insects almost certainly corresponds with the may-fly's sub-imago, and the facts just recalled as to remnants of pupal activity suggest that in the ancestors of endopterygote insects what is now the pupal instar was represented by an active nymphal or sub-imaginal stage, possibly indeed by more than one stage, as Packard and other writers have stated that pupae of bees and wasps undergo two or three moults before the final exposure of the imago. Such an early pupal instar has been defined as a 'pro-nymph' or a 'semi-pupa.' Examples have been given of the exceptional passive condition of the penultimate instar in Exopterygota. The instars preceding this presumably had originally outward wing-rudiments in all insect life-histories, and the endopterygote condition was attained by the postponement of the outward appearance of these to successively later stages. The leg and wing rudiments of the male coccid (pp. 20-1) beneath the cuticle of the second instar are strictly comparable to imaginal buds, and these are present in one instar of what is generally regarded as an exopterygote life-history. The first instar in all insects has no visible wing-rudiments, but when they grow outwardly from the body, they necessarily become covered with cuticle, so that they must be visible after the first moult. There is no supreme difficulty in supposing that the important change was for these early rudiments to become sunk into the body, so that the cuticle of the second, and, later, of the third and succeeding instars, showed no outward sign of their presence. This suggestion is confirmed by Heymons' (1896, 1907) observation of the occasional appearance of outward wing-rudiments on the thoracic segments of a mealworm, the larva of the beetle Tenebrio molitor, and by F. Silvestri's discovery (1905) of a 'pro-nymph' stage with short external wing-rudiments between the second larval and the pupal instars of the small ground-beetle Lebia scapularis. Whatever may be the exact explanation of these abnormalities, they show that in the life-story of the higher insects outward wing-rudiments may even yet appear before the pupal stage, confirming our belief that such appearance is an ancestral character. The inward growth of these wing-rudiments may well have been correlated with a difference in form between the newly-hatched insect and its parent. As this difference persisted until a constantly later stage, and the pre-imaginal instar became necessarily a stage for reconstruction, the present condition of complete metamorphosis in the more highly organised orders was finally attained.

To explain satisfactorily these complex life-stories is however admittedly a difficult task. The acquisition of wings is, as we have seen, a dominating feature in them all, but if we try to go yet a step farther back and speculate on the origin of wings in the most primitive exopterygote insects, the task becomes still more difficult. Many years ago Gegenbaur (1878) was struck by the correspondence of insect wings to the tracheal gills of may-fly larvae, which are carried on the abdominal segments somewhat as wings are on the thoracic segments. But B鍵ner has recently (1909) brought forward evidence that these abdominal gills really correspond serially with legs. Moreover Gegenbaur's theory suggests that the ancestral insects were aquatic, whereas the presence of tubes for breathing atmospheric air in well-nigh all members of the class, and the fact that aquatic adaptations, respiratory and otherwise, in insect-larvae are secondary force the student to regard the ancestral insects as terrestrial. It is indeed highly probable that insects had a common origin with aquatic Crustacea, but all the evidence points to the ancestors of insects having become breathers of atmospheric air before they acquired wings. How the wings arose, what function their precursors performed before they became capable of supporting flight, we can hardly even guess.

Our study of the life-story of insects, therefore, while it has taught us something of what is going on around us to-day, and has given us hints of the course of a few threads of that long life-story which runs through the ages, brings us face to face with the most instructive, if humbling fact that 'there are many more things of which we are ignorant.' The passage from creeping to flight, as the caterpillar becomes transformed into the butterfly, was a mystery to those who first observed it, and many of its aspects remain mysterious still. Perhaps the most striking result of the study of insect transformation is the appreciation of the divergent specialisation of larva and imago, and it is a suggestive thought that of the two the larva has in many cases diverged the more from the typical condition. The caterpillar crawling over the leaf, or the fly-grub swimming through the water, may thus be regarded as a creature preparing for a change to the true conditions of its life. It is a strange irony that the preparation is often far longer than the brief hours of achievement. But the light which research has thrown on the nature of these wonderful life-stories, the demonstration of the unseen presence and growth within the insect, during its time of preparation among strange surroundings, of the organs required for service in the coming life amid its native air, confirm surely the intuition of the old-time students, who saw in these changes, so familiar and yet so wonderful, a parable and a prophecy of the higher nature of man.

OUTLINE CLASSIFICATION OF INSECTS

Class INSECTA or HEXAPODA.

Sub-class A, APTERYGOTA.

Order 1. Thysanura (Bristle-tails). 2. Collembola (Spring-tails).

Sub-class B, EXOPTERYGOTA.

Order 1. Dermaptera (Earwigs). 2. Orthoptera (Cockroaches, Grasshoppers, Crickets). 3. Plecoptera (Stone-flies). 4. Isoptera (Termites or 'White Ants'). 5. Corrodentia (_a_) Copeognatha (Book-lice). (_b_) Mallophaga (Biting-lice). 6. Ephemeroptera (May-flies). 7. Odonata (Dragon-flies). 8. Thysanoptera (Thrips). 9. Hemiptera (_a_) Heteroptera (Bugs, Pond-skaters) (_b_) Homoptera (Cicads, 'Greenfly,' Scales). 10. Anoplura (Lice).

Sub-class C, ENDOPTERYGOTA.

Order 1. Neuroptera (Alder-flies, Ant-lions, Lacewings). 2. Coleoptera (Beetles). 3. Mecaptera (Scorpion-flies). 4. Trichoptera (Caddis-flies). 5. Lepidoptera (Moths and Butterflies). 6. Diptera (Two-winged flies) (_a_) Orthorrhapha (Crane-flies, Midges, Gnats) (_b_) Cyclorrhapha (Hover-flies, House-flies, Bot-flies, &c). 7. Siphonaptera (Fleas). 8. Hymenoptera (_a_) Symphyta (Saw-flies) (_b_) Apocrita (Gall-flies, Ichneumon-flies, Wasps, Bees, Ants).

TABLE OF GEOLOGICAL SYSTEMS

These names, given by geologists to the various divisions of rocks, as indicated by the fossils entombed in them, are arranged in 'descending' order, the more recent formations above, the more ancient below, as newer deposits necessarily lie over older beds.

CALNOZOIC OR TERTIARY GROUP.

Pleistocene. Pliocene. Miocene. Eocene.

MESOZOIC OR SECONDARY GROUP.

Cretaceous. Jurassic. Triassic.

PALAEOZOIC OR PRIMARY GROUP.

Permian. Carboniferous. Devonian. Silurian. Cambrian.

BIBLIOGRAPHY

The following list of some books and papers, referred to in this little volume or of especial service to the author in its preparation, is needless to say very far from exhaustive. To save space, titles are often abbreviated. Most of the works in the general list (A) contain extensive lists of literature on insects and their transformations, these should be consulted by the serious student.

A. GENERAL WORKS.

1909. C. Boner. Die Verwandlungen der Insekten. _Sitzb. d. Gesellsch. naturforsch. Freunde, Berlin._

1869. F. Brauer. Betrachtung die Verwandlung der Insekten. _Verhandl. der K.K. zool.-bot. Gesellschaft in Wien._ XIX.

1899. G.H. Carpenter. Insects, their Structure and Life. London.

1859. C. Darwin. The Origin of Species. London.

1909. P. Deegener. Die Metamorphose der Insekten. Leipzig.

1906. J.W. Folsom. Entomology. London.

1878. C. Gegenbaur. Grundriss der Vergleichende Anatomie. Leipzig.

1906. A. Handlirsch. Die fossilen Insekten. Leipzig.

1904. L.F. Henneguy. Les Insectes. Paris.

1907. R. Heymons. Die verschiedenen Formen der Insectenmetamorphose. _Ergebnisse der Zoologie._ I.

1899. A. Lameere. La raison d'être re des Metamorphoses chez les Insectes. _Ann. Soc. Entom. Bruxelles._ XLIII.

1874. J. Lubbock. The Origin and Metamorphoses of Insects. London.

1895. L.C. Miall. (_a_) The Transformations of Insects. _Nature._ LIII.

1895. ---- (_b_) The Natural History of Aquatic Insects. London.

1908. ---- Injurious and Useful Insects. 2nd edition. London.

1839. G. Newport. Insects. _Todd Cyclopaedia._ II. London.

1898. A.S. Packard. Text book of Entomology. New York.

1734-42. R.A.F. de Rumur. Memoires pour servir ?l'Histoire naturelle et ?l'anatomie des Insectes. Paris.

1895-8. D. Sharp. The Cambridge Natural History, V, VI. London.

1899. ---- Some points in the Classification of Insects. IV. _Internat. Zoolog. Congress._

1902. ---- Insects in _Encycl. Brit._ 10th Edition, XXIX. London.

1910. ---- and G.H. Carpenter. Hexapoda in _Encycl. Brit._ 11th Edition. Cambridge.

1737. J. Swammerdam. Biblia Naturae. Leyden (incorporates works on Insects published during the author's lifetime 1669-75).

1909. F.V. Theobald. Insect Pests of Fruit. Wye.

B. SPECIAL WORKS.

1881. H. Adler. Ueber den Generationswechsel den Eichen-Gallwespen. _Zeitsch. f. wissensch. Zoologie._ XXXV.

1896. ---- and C.R. Straton. Alternating Generations. Oxford.

1902. J. Anglas. Nouvelles Observations sur les Metamorphoses Internes. _Arch. d'Anat. Microscop._ IV.

1911. E.E. Austen. Handbook of the Tsetse-Flies. London (Brit. Museum).

1909. F. Balfour-Browne. Life-History of Agrionid Dragonfly. _Proc. Zool. Soc. Lond._

1893, &c. C.G. Barrett. Lepidoptera of the British Islands. London.

1890. H. Beauregard. Les Insectes Vicants. Paris.

1909. C. Boner. Die Tracheenkiemen der Ephemeriden. _Zoolog. Anz._ xxxiii.

1863. F. Brauer. Monographie der Oestriden. Wien.

1894. C. Brongniart. Rherches pour servir ?l'histoire des Insectes fossiles des Temps Primaires. St Etienne.

1893. T.A. Chapman. Structure of Pupae of Heterocerous Lepidoptera. _Trans. Entom. Soc. Lond._

1891. H. Dewitz. Das geschlossene Tracheensystem bei Insektenlarven. _Zoolog. Anz._ xiii.

1857-8. J.H. Fabre. L'Hypermamorphose et les Moeurs des Meloides. _Ann. Sci. Nat._ (_Zool._), (4). VII. IX.

1869. M. Ganin. Die Entwicklungsgeschichte bei den Insekten. _Zeitsch. f. wissensch. Zoolog._ xix.

1894. J. Gonin. La Metamorphose des Lepidopteres. _Bull. Soc. Vaud. Sci. Nat._ xxx.

1870. O. Grimm. Die ungeschechtliche Fortpflanzung einer Chironomus. _Mem. Acad. Imp. St Petersbourg_ (7). xv.

1890. W. Hatchett-Jackson. Morphology of the Lepidoptera. _Trans. Linn. Soc. (Zool.) Lond._ (2). v.

1896. R. Heymons. Flelbildung bei der Larve von Tenebrio molitor. _Sitzb. d, Gesellsch. Naturforsch. Freunde, Berlin._

1906. ---- Ueber die ersten Jugendformen von Machilis alternata. _Ib._

1908. W. Kahle. Die Paedogenesis der Cecidomyiden. _Zoologica._ IV.

1913. V.L. Kellogg. Distribution and Species-forming of Ectoparasites. _Amer. Naturalist._ XLVII.

1887. A. Kowalevsky. Die nachembryonale Entwicklung der Musciden. _Zeitsch. f. wissensch. Zool._ XLV.

1904. O.H. Latter. Natural History of Common Animals (chaps. III, IV, V). Cambridge.

1890-95. B.T. Lowne. The Blowfly, 2 vols. London.

1863. J. Lubbock. Development of Chloeon. _Trans. Linn. Soc. Lond._ XXIII.

1762. P. Lyonet. Trait?anatomique de la Chenille. Haag.

1669. M. Malpighi. De Bombyce. London.

1898. C.L. Marlatt. The periodical Cicada. _Entom. Bull._ 14, _U.S. Dept. Agric._

1898. G.A.K. Marshall. Seasonal Dimorphism in Butterflies. _Ann. Mag. Nat. Hist._ (7). II.

1900. L.C. Miall and A.B. Hammond. The Harlequin Fly. Oxford.

1901-3. R. Newstead. Coccidae of the British Isles. London.

1877. J.A. Palm 闓. Zur Morphologie des Tracheensystems. Leipzig.

1891. E.B. Poulton. External Morphology of the Lepidopterous Pupa. _Trans. Linn. Soc. Zool._ (2). V.

1892. ---- Colour-relation between Lepidopterous Larvae &c. and their surroundings. _Trans. Entom. Soc. Lond._

1880. C.V. Riley. Pupation of Butterflies. _Proc. Amer. Assoc._ XXVIII.

1902. E.D. Sanderson. Report of Entomologist. Delaware. U.S.A.

1885. E.O. Schmidt. Metamorphose und Anatomie des m 銕 nlichen Aspidiotus. _Archiv f. Naturgeschichte._ LI.

1885. S.H. Scudder. Insekten in Zittel's Paleontologie. II.

1907. A.J. Siltala. Die postembryonale Entwicklung der Trichopteren-Larven. _Zoolog. Jahrb. Suppl._ IX.

1905. F. Silvestri. Metamorfosi e Costumi della Lebia scapularis. _Redia._ II.

1900. J.B. Smith. The Apple Plant-louse. _New Jersey Agric. Exp. Station Bull._ 143.

1888. J. Van Rees. Die innere Metamorphose von Musca. _Zoolog. Jahrb. Anat._ III.

1911. K.W. Verhoeff. Ueber Felsenspringer, Machiloidea. _Zoolog. Anz._ XXXVIII.

1865. N. Wagner. Die viviparen Gallm 點 kenlarven. _Zeitsch. f. wissensch. Zoolog._ XV.

1901. E. Wasmann. Termitoxenia. _Zeitsch. f. wissensch. Zoolog._ LXX.

1864. A. Weismann. Die nachembryonale Entwicklung der Musciden. _Zeitsch. f. wissensch. Zoolog._ XIV.

1865. ---- Die Metamorphose von Corethra. _Ib._ XVI.

1876. ---- Studien zur Descendenz-Theorie. Leipzig. (English Translation by R. Meldola, London, 1882.)